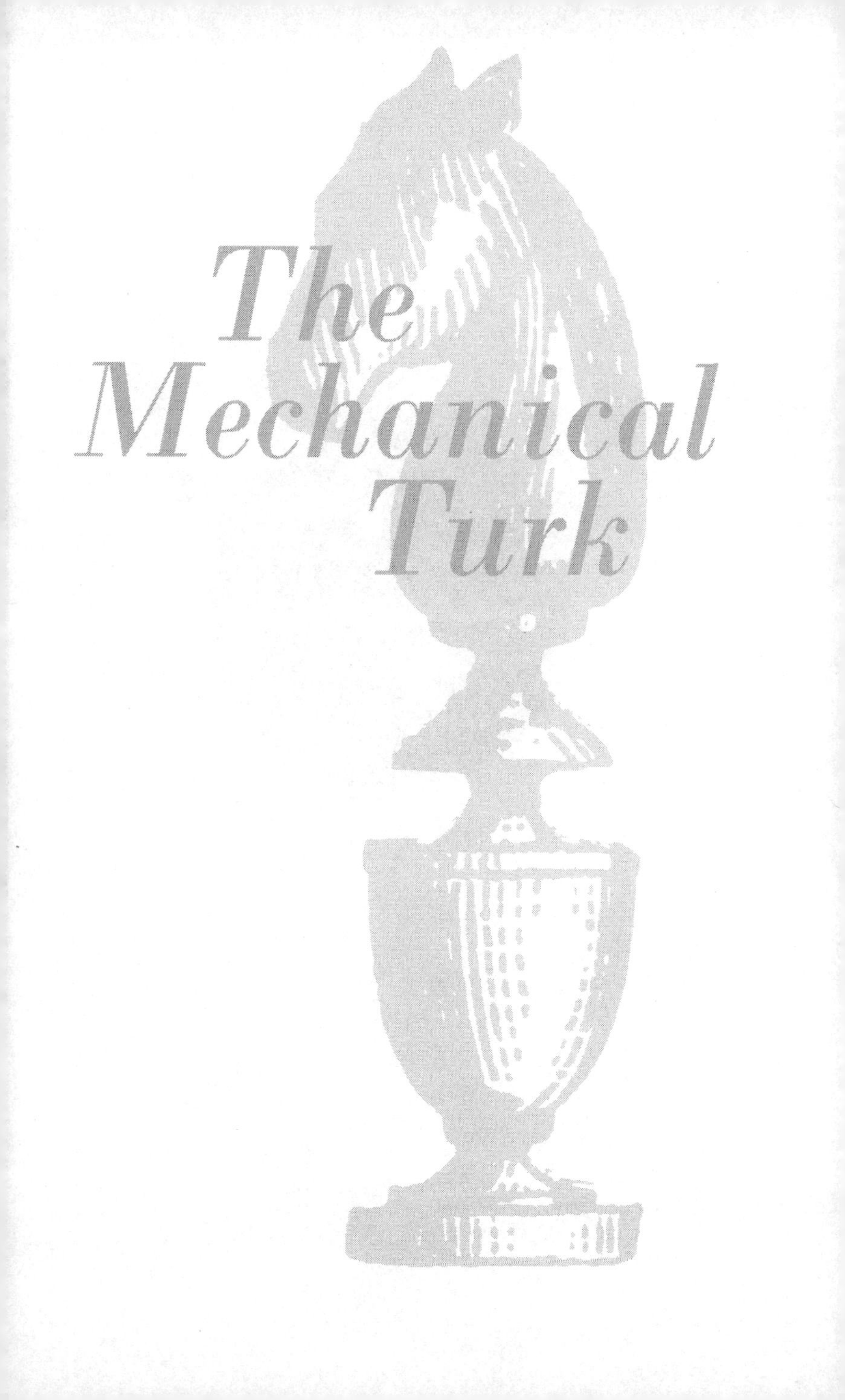
The Mechanical Turk

The Mechanical Turk

The True Story of
the Chess-Playing Machine
That Fooled the World

Tom Standage

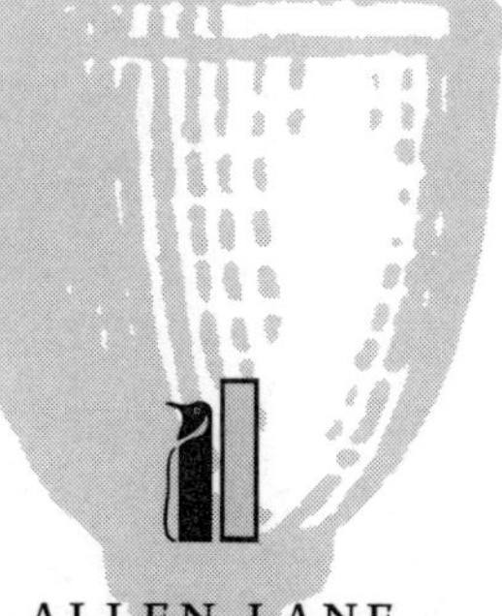

ALLEN LANE
THE PENGUIN PRESS

ALLEN LANE
THE PENGUIN PRESS

Published by the Penguin Group
Penguin Books Ltd, 80 Strand, London WC2R ORL, England
Penguin Putnam Inc., 375 Hudson Street, New York, New York 10014, USA
Penguin Books Australia Ltd, 250 Camberwell Road, Camberwell, Victoria 3124, Australia
Penguin Books Canada Ltd, 10 Alcorn Avenue, Toronto, Ontario, Canada M4V 3B2
Penguin Books India (P) Ltd, 11, Community Centre, Panchsheel Park, New Delhi – 110 017, India
Penguin Books (NZ) Ltd, Cnr Rosedale and Airborne Roads, Albany, Auckland, New Zealand
Penguin Books (South Africa) (Pty) Ltd, 24 Sturdee Avenue, Rosebank 2196, South Africa

Penguin Books Ltd, Registered Offices: 80 Strand, London WC2R ORL, England

www.penguin.com

First published in the USA as *The Turk* by Walker Publishing Co.
and simultaneously in Canada by Fitzhenry and Whitside 2002
First published in Great Britain under the present title by
Allen Lane The Penguin Press 2002

1

Set in 11.75/15.5 pt PostScript Monotype Dante
Printed in England by Clays Ltd, St Ives plc

ISBN 0-713-99525-4

For Ella

Contents

Contents

The Mechanical Turk

Preface

Automaton (from αυτοζ, self, and μαω, to seize): a self-moving machine, or one in which the principle of motion is contained within the mechanism itself. According to this description, clocks, watches and all machines of a similar kind are automata, but the word is generally applied to contrivances which simulate for a time the motions of animal life.

—*Encyclopædia Britannica*, 11th edition (1911)

On an autumn day in 1769, Wolfgang von Kempelen, a thirty-five-year-old Hungarian civil servant, was summoned to the imperial court in Vienna by Maria Theresa, empress of Austria-Hungary, to witness the performance of a visiting French conjuror. Kempelen was well versed in physics, mechanics, and hydraulics, and was a trusted servant of the empress. She had invited him on a whim because she wanted to see what an expert in scientific matters would make of the conjuror's tricks. Yet the performance was to change the course of Kempelen's life. It set in motion a chain of events that led him to construct an extraordinary machine: a mechanical man, dressed in an oriental costume, seated behind a wooden cabinet, and capable of playing chess.

At the time, elaborate mechanical toys were a popular form of entertainment in the courts of Europe, though the technology they embodied was soon to be put to more serious uses. So Kempelen intended his chess-playing machine to do little more than amuse the court and advance his career by impressing the empress. But instead his automaton unexpectedly went on to achieve widespread fame throughout Europe and America, bringing Kempelen both triumph and despair. During its eighty-five-year career the automaton was associated with a host of historical figures, including Benjamin Franklin, Catherine the Great, Napoleon Bonaparte, Charles Babbage, and Edgar Allan Poe. It was the subject of numerous stories and anecdotes and inspired many legends and outright fabrications, the truth of many of which will never be known. The chess player was, in fact, destined to become the most famous automaton in history. And along the way, Kempelen's work would unwittingly help to inspire the development of the power loom, the telephone, the computer, and the detective story.

To modern eyes, in an era when it takes a supercomputer to defeat the world chess champion, it seems obvious that Kempelen's chess-playing machine had to have been a hoax – not a true automaton at all but a contraption acting under the surreptitious control of a human operator, like a puppet dancing on a string. How, after all, would it have been possible to build a genuine chess-playing machine using eighteenth-century clockwork and mechanical technology? But during the eighteenth century automata of extraordinary ingenuity were being constructed and exhibited across Europe, including Jacques de Vaucanson's mechanical duck, Henri-Louis Jaquet-Droz's harpsichord player, and John Joseph Merlin's dancing lady. Mechanical devices seemed to offer limitless new technological possibilities. So the notion that Kempelen's machine really could play chess did not seem totally out of the question.

Even among the sceptics who insisted it was a trick, there was disagreement about how the automaton worked, leading to a series of claims and counterclaims. Did it rely on mechanical trickery, magnetism, or sleight of hand? Was there a dwarf, or a small child, or a legless man hidden inside it? Was it controlled by a remote operator in another room or concealed under the floor? None of the many explanations put forward over the years succeeded in fully fathoming Kempelen's secret and served only to undermine each other. Indeed, it is only recently, following the construction of a replica of the automaton, that the full secret of its operation has been uncovered.

By choosing to make his machine a chess player, a

contraption apparently capable of reason, Kempelen sparked a vigorous debate about the extent to which machines could emulate or replicate human faculties. The machine's debut coincided with the beginnings of the industrial revolution, when machines first began to displace human workers, and the relationship between people and machines was being redefined. The chess player posed a challenge to anyone who took refuge in the idea that machines might be able to outperform humans physically but could not outdo them mentally. The reactions it inspired thus foreshadowed modern reactions to the computer, over 200 years later. And the automaton's curious tale, running in a parallel course alongside the prehistory of computing but connecting in a few key places, has now assumed a new significance as scientists and philosophers continue to debate the possibility of machine intelligence.

Kempelen never gave his automaton a name, but its distinctive oriental costume gave rise to a nickname almost immediately, and it is known to this day as the Turk. This is the story of its remarkable and chequered career.

CHAPTER ONE

The Queen's Gambit Accepted

THE QUEEN'S GAMBIT (D4 D5 C4): An opening in which White attempts to sacrifice his queen's bishop's pawn to accelerate the development of his position. Black accepts the gambit by taking the offered pawn.

You seek for knowledge and wisdom as I once did; and I ardently hope that the gratification of your wishes may not be a serpent to sting you, as mine has been.

– Mary Shelley, *Frankenstein* (1818)

Automata are the forgotten ancestors of almost all modern technology. From computers to compact-disc players, railway engines to robots, the origins of today's machines can be traced back to the elaborate mechanical toys that flourished in the eighteenth century. As the first complex machines produced by man, automata represented a proving ground for technology that would later be harnessed in the industrial revolution. But their original uses were rather less utilitarian. Automata were the playthings of royalty, both as a form of entertainment in palaces and courts across Europe and as gifts sent from one ruling family to another. As well as being a source of amusement, automata provided a showcase for each nation's scientific prowess, since they embodied what was, at the time, the absolute cutting edge of new technology. As a result, automata had a far greater social and cultural importance than their outward appearance as mere toys might suggest.

The first automata were essentially scaled-down versions of the elaborate mechanical clocks that adorned cathedrals across Europe from medieval times. As well as displaying the time, these clocks often had astronomical features (such as the phase of the moon) and, in some cases, entire mechanical theatres that sprang to life on particular occasions. A typical configuration involved figures of the Madonna and Child, who would appear through a doorway on specific feast days as the clock struck the hour. They would be followed by figures representing the three kings, shepherds, and so on, all of whom would genuflect before the Madonna, present their gifts, and then disappear through another door. A good example can still be seen today on the clock tower of St. Mark's in Venice. Municipal clocks in town squares subsequently adapted this formula but replaced the religious figures with kings, knights, trumpeters, birds, and other animals. These clocks provided the inspiration for smaller and increasingly elaborate automata that clockmakers sold to rich customers. As these devices became more complicated, their time-keeping function became less important, and automata became first and foremost mechanical amusements in the form of mechanical theatres or moving scenes.

One popular kind of automaton was the mechanical picture, a painting with moving parts driven by an elaborate clockwork mechanism hidden behind or within the frame. Another type of automaton, also intended as a conversation piece, took the form of a table ornament. Such devices could hold cutlery, napkins, and spices, had spouts to dispense wine or water, were decorated with moving figures or animals, and often incorporated a clock. A particularly fine example, made for Emperor Rudolph II by Hans Schlottheim, a German automaton maker, can be seen today in the British Museum.

Another influence on the design of automata was the long tradition of imitating nature through the construction of mechanical animals. The Italian artist and inventor Leonardo da Vinci, for example, designed a flying machine modeled on a bird and is said to have made a mechanical lion. His fifteenth-century German contemporary, Johann Müller, known as "Regiomontanus", presented Emperor Maximilian with an iron fly and a mechanical eagle, which is reputed to have escorted the emperor to the city gates of Nuremberg, though exactly how is unclear. Even less plausible is the brass fly constructed by Bishop Virgilius of Naples. It supposedly chased all the real flies from the city, which remained free of flies for eight years.

Inspired by such tales, makers of automata enjoyed the challenge of making machines that were capable of moving in a lifelike manner. There were music boxes and snuffboxes out of which singing birds or dancing figures appeared, and innumerable mechanical animals. One eighteenth-century automaton-maker, an Englishman named James Cox, made an eight-foot-high mechanical elephant encrusted with diamonds, rubies, emeralds, and pearls. Cox was renowned for his automata and mechanical clocks, many of which were sold or sent as gifts to China by the East India Company. His other creations included a mechanical tiger, a peacock, and a swan.

Sometimes automata imitated living things a little too credibly, as was the case with a supposed automaton harpsichord player that made an appearance at the court of the French king Louis XV during the 1730s and enchanted listeners with its musical ability. The king insisted on being shown the mechanism that could play in such a charming and lifelike manner, whereupon a five-year-old girl was found concealed inside the machine.

Other famous (but genuine) automata included the writer, draftsman, and harpsichord player constructed by Henri-Louis Jaquet-Droz, a member of a Swiss family of clockmakers. The movements of these automata, which could write, draw, and play music respectively, were programmed using irregularly shaped discs, called cams, threaded onto a spindle. As the spindle rotated, spring-loaded levers resting on the cams moved up and down, and controlled the motion of the automaton's various parts by pushing and pulling on connecting rods. By paying meticulous attention to the shapes of the various cams, one could programme an automaton to make coordinated, lifelike movements of extraordinary grace and subtlety. Similar writing automata were built in the 1750s for Maria Theresa, empress of Austria-Hungary, by Friedrich von Knauss, an Austrian inventor who is also credited with the invention of the typewriter.

Since only the very rich could afford to buy their extravagant contraptions, makers of automata moved in elevated circles and often ended up in the direct employ of kings, queens, and emperors. Building automata thus provided a good way for serious-minded clockmakers, engineers, or scientists seeking patronage to demonstrate their abilities and establish reputations for themselves; tinkering with mechanical toys could lead to both fame and fortune. Perhaps the best example is provided by the Frenchman Jacques de Vaucanson, whose inventions dazzled Europe in the mid-eighteenth century, and whose renown as an automaton maker enabled him to move effortlessly between the worlds of entertainment, industry, and science.

*

Vaucanson was born in 1709, the youngest of ten children, and studied theology at the Jesuit college in Grenoble with a view to becoming a monk. He also enjoyed building mechanical toys, and he soon found that this was incompatible with his religious vocation. According to one story, he built tiny flying toys in the form of angels, which angered his superiors; another tale suggests that it was a table automaton that got Vaucanson into trouble with a senior official of his religious order. In any case, forced to choose between his religious calling and his enthusiasm for elaborate machinery, he renounced the religious life and decided instead to devote himself to building automata.

Like other automaton makers, Vaucanson was particularly interested in building machines capable of imitating the natural processes of living beings, including respiration, digestion, and the circulation of the blood. His ultimate goal was to build an artificial man. But Vaucanson soon realized that in order to pursue this goal, he would first have to put his talents to commercial use and raise money "by producing some machines that could excite public curiosity". Displays of automata were becoming increasingly popular, particularly in Paris and London, where they provided an opportunity for the public to witness a variety of elaborate machinery that they would never have been able to afford to buy for themselves.

The automaton that first brought Vaucanson to public attention took the form of a flute player. One day in 1735, while walking through some public gardens in Paris, he saw a statue of a boy holding a flute to his lips and was inspired to build a moving statue that could actually play melodies. The primary purpose of the automaton was to enable Vaucanson to investigate the human respiratory system, and to this end he furnished it with artificial lungs, windpipe, and mouth, to which it held its flute. The lungs consisted of three sets of bellows, driven by a rotating crankshaft, to ensure a constant flow of air at low, medium, and high pressure. A set of valves adjusted the amount of air at each pressure that was allowed into the windpipe, and another valve in the mouth regulated the airflow, performing the function of the tongue. The movements of these valves, together with those of the fingers and the lips, were controlled by a set of spring-loaded levers whose ends rested on the surface of a rotating drum. The surface of the drum was covered with small studs; as the ends of the levers passed over these studs, they rose and fell, causing the automaton to move its fingers and lips accordingly. This meant that every aspect of the automaton's complex operation could be programmed in advance by inserting a suitable configuration of studs into the surface of the drum. The automaton could thus be made to play intricate melodies and mimic almost all of the subtleties of a human flute player's breathing and musical expression.

Vaucanson put his flute player on public display in Paris in October 1737, and it was an immediate success. Mindful of the false automaton that had deceived the court of Louis XV, Vaucanson subsequently allowed his flute player to be scrutinized by members of the Academy of Sciences in Paris, one the world's leading scientific societies, to dispel any question of trickery. One account of the event written by Juvigny, a French politician, recorded that "at first many people would not believe that the sounds were produced by the flute which the automaton was holding. These people believed that the sounds must come from an organ enclosed in the body of the figure. The most incredulous, however, were soon convinced that the automaton was in fact blowing the flute, and that the breath coming from his lips made it play and that the movement of his fingers determined the different notes. The machine was submitted to the most minute examination and to the strictest tests. The spectators were permitted to see even the innermost springs and to follow their movements." Vaucanson's flute player was thus proven to be an entirely genuine automaton. What the false automaton had accomplished through trickery, Vaucanson had achieved through a combination of ingenuity and the latest in mechanical technology.

Within a few months he had completed a second automaton, this time of a boy playing a pipe with one hand and a drum with the other. With only one hand to play the three-holed pipe, the sound it produced was far more dependent on the air pressure, the tonguing, and the position of the automaton's fingers. It thus presented a further challenge to Vaucanson's ability to mimic human subtleties. But it was Vaucanson's third automaton, a model of the digestive system, that was to become his most famous creation. Instead of building it in the form of a person, Vaucanson decided to imitate an animal and built a mechanical duck.

He described this automaton in a letter to a contemporary as "an artificial duck made of gilded copper that drinks, eats, quacks, splashes about on the water, and digests his food like a living duck". The duck could stretch out its neck, take grain from a spectator's hand, and then swallow, digest, and excrete it. The duck's wings were anatomically exact copies of real wings, with each bone rendered in metal and adorned with a few feathers. The duck could even flap its wings and create a gentle breeze. But while spectators were chiefly struck by the extraordinarily lifelike nature of the duck, Vaucanson was chiefly interested in its innards, which he left exposed to view. The duck's insides imitated the digestive process by dissolving the grain in an artificial stomach, from where it was passed along tubes and excreted. In the process of building this automaton, Vaucanson pioneered the development of flexible rubber tubing.

In common with Vaucanson's other automata, the duck was mounted on a wooden pedestal, and its mechanism was powered by a falling weight, in the same way as a grandfather clock. The weight was suspended on a cord, which was wrapped around a large drum. As the weight fell, it turned the drum, thus directing the duck's movements through an elaborate system of cams and levers. In the words of Juvigny, "During the time that this artificial animal was eating grain from someone's hand, drinking and splashing in the water brought to him in a vase, passing his excrements, flapping and spreading his wings and imitating all the movements of a living duck, everybody was allowed to look inside the pedestal. In this were all the wheels, all the levers, and all the wires communicating through the animal's legs with the different parts of his body and this was likewise open to view. As with the flute-player, a weight was the one and only source of power to set the whole thing in motion and keep it moving."

A Lithograph by Albert Chereau showing Vaucanson's automata: the flute player and the mechanical duck.

Such was the acclaim that greeted these extraordinary machines – Voltaire described their inventor as "bold Vaucanson, rival to Prometheus" – that Vaucanson allowed them to go on a tour of the courts of Europe, as ambassadors for French ingenuity and scientific advancement. Vaucanson was made a member of the Academy of Sciences in Paris; King Frederick II of Prussia offered him a job with a generous salary of 12,000 livres; he was even given the opportunity by Louis XV of France to go on an expedition to Guiana in order to further the development of his new rubber tubing.

However, Vaucanson decided to stay in France and pursue his goal of building an artificial man. Once it was completed, he hoped to use this automaton "to perform experiments on animal functions, and thence to gather inductions to know the different states of health of men so as to remedy their ills". But this ambitious project quickly stalled, so in 1741 Vaucanson accepted the offer of the lucrative government post of inspector of manufactures, with responsibility for applying his mechanical ingenuity to the modernization of the French weaving industry. He drew up elaborate plans to transform manufacturing methods and work practices. But his reorganization plans were abandoned when the silk workers of the city of Lyons, who were to try out his new ideas, heard of his scheme and complained that they would be herded into factories and forced to act as mere drudges on a production line. Wary of becoming human parts in what would be, in effect, a huge automaton, they rioted in the streets, forcing Vaucanson to disguise himself as a monk and flee for his life.

Vaucanson returned to Paris, where he decided to withdraw from the limelight. In 1743, he sold his trio of automata to a consortium of businessmen from Lyons, who showcased them at the Haymarket theatre in London and subsequently displayed them across Europe. Vaucanson was appointed official examiner of new machine inventions at the Academy of Sciences in Paris and spent his remaining years working on many other inventions, including a number of improvements to machine tools such as lathes, milling machines, and drills. He also devised a machine to manufacture an endless chain and spent many years working on a power loom that could weave silk automatically, without the need for human intervention. With this machine, Vaucanson declared, "a horse, an ox or an ass can make cloth more beautiful and much more perfect than the most able silkworkers . . . each machine makes each day as much material as the best worker, when he is not wasting time." But his weaving machine never got past the experimental stage and was not adopted by the weaving industry. Vaucanson never built his artificial man either. He was, however, responsible for causing a surge in public interest in automata. His work paved the way for many subsequent inventions and inspired other automaton makers – including Wolfgang von Kempelen.

*

As one of Maria Theresa's senior officials, Wolfgang von Kempelen would have seen a procession of automata and other scientific amusements being presented to the empress at her court in Vienna, including musical automata, mechanical animals, and other contraptions. But he was no ordinary observer, for he had taught himself the principles of physics, mechanics, and hydraulics, even though he had come to the subject relatively late in life. This meant he was able to appreciate how the various automata worked, and to observe which ones were regarded as most impressive by spectators. At some point, he started to hatch a plan for an automaton of his own.

As a wealthy civil servant, Kempelen was an unlikely automaton maker; it seems he was simply looking for a challenge beyond the humdrum routine of his day-to-day duties. For although he was doing well in his career, life at the court was insufficiently stimulating to someone with such a wide range of interests.

Born in 1734, as a young man Kempelen had studied philosophy and law in Vienna. He then made an artistic pilgrimage to Italy before being formally introduced to the Viennese court by his father, Engelbert, a retired customs officer, in 1755. A strikingly handsome twenty-one-year-old

Wolfgang von Kempelen

who spoke several languages, Kempelen made an immediate impression. He was given the important task of translating the Hungarian civil code from Latin into German, which Maria Theresa had made the official language throughout her newly united kingdom of Austria-Hungary. Kempelen retired to his living quarters and completed the work in a few days. His translation was hailed as a masterpiece; it seemed extraordinary that he could have produced so flawless a translation of such a complex text in so little time. Kempelen was soon appointed counsellor to the imperial court, with a salary three times what his father had earned. On the official document confirming his appointment, Maria Theresa wrote, "The Hungarian court will benefit greatly from young Mr Kempelen."

Kempelen was indeed a valuable asset to the court: he was hardworking and conscientious in his professional capacity, while being charming and gregarious in person. In September 1757, with his fortunes rising fast, Kempelen married a lady-in-waiting at the court, and soon afterwards he was promoted further. But Kempelen's wife, Franciscka, died suddenly a few weeks later. Shocked and grief-stricken, Kempelen responded by immersing himself in his hobby: scientific investigation. As a wealthy man, he was able to afford the expensive materials needed to equip his own workshop, where he devoted his spare time to research and experimentation. He swiftly collected an assortment of the latest scientific equipment and all the wood- and metal-working tools of a joiner, a locksmith, and a watchmaker. Adjoining his workshop was his study, which was lined with books, antiques, and engravings. One of Kempelen's friends wrote of him that "his predominant passion is invention, in which he employs almost every moment which the duties of his situation leave at his disposal."

As his interest in science and mechanics grew, Kempelen continued to prosper at the court. In 1758 he was appointed controller of the imperial salt mines in Transylvania, and he was promoted to director of the mines in 1766, by which time he had also remarried. He now felt confident enough to put his scientific knowledge into practice, and he devised a system of pumps to drain the mines when they became flooded with water. Following the success of this project, he was asked to design the waterworks for the castle in his hometown of Pressburg, the capital of Hungary, a few miles to the east of Vienna. (Pressburg was called Poszony in Hungarian and is now the Slovakian town of Bratislava.)

In 1768 Kempelen was given the challenging task of coordinating the settlement of the mountainous Banat province of Hungary. While in Banat he solved a local mystery, freeing several wrongly imprisoned men from jail. He also planned villages and designed houses, and over the next three years thousands of families settled in the region. During this time Kempelen spent a lot of time in Banat but made frequent visits to Vienna to report back on his progress. It was on one of these visits, in the autumn of 1769, that he was invited by Maria Theresa to attend the scientific conjuring show being presented to the court that evening by a visiting Frenchman named Pelletier.

Maria Theresa was particularly interested in science and had an unusually enlightened attitude towards it for her time. Soon after coming to the throne she had, for example, taken a strong line against the overzealous persecution of people accused of being vampires or witches. On one occasion she pardoned a man who had been found guilty of witchcraft and was due to be beheaded, declaring, "Witches can only be found where there is ignorance. This man is no more capable of witchcraft than I." She was also an advocate of the practice of inoculation against smallpox. Following an outbreak of the disease in Vienna in 1767 that claimed

Maria Theresa

the lives of several members of her own family, the empress had her own sons inoculated and subsequently paid for the inoculation of dozens of poor children.

The empress was aware of Kempelen's growing reputation in scientific circles and hoped he would be able to explain to her how Pelletier's conjuring tricks worked. Kempelen was known to be good at explaining technical matters when asked to do so, without being a bore. "It is very rare to hear him speak of mechanism, notwithstanding it is his dominant passion," noted one of his friends, who praised the "astonishing fluency" of Kempelen's explanations "if the conversation be led to this subject". Kempelen agreed to do his best to explain the conjurer's tricks to the empress and took a seat near her in the audience. Pelletier, who is thought to have been a member of the prestigious Academy of Sciences in Paris, finished preparing his equipment and indicated that he was ready to begin. Maria Theresa nodded, and the performance began.

*

The exact nature of the "magnetic games" performed by Pelletier is uncertain. But his routine probably had as much in common with a scientific lecture as with a modern conjuring show. It is likely that there would have been chemical reactions, explosions, demonstrations of magnetism, and a number of tricks involving automata. Mixing scientific demonstrations and automata with more traditional, oldfangled conjuring would have given the whole performance a vital veneer of scientific respectability. At the time, conjurors were at pains to stress that their tricks relied on "natural" (or "white") magic and thus did not contravene the divine laws of nature, unlike "supernatural" (or "black") magic, which was thought to involve the intervention of the devil.

Throughout the performance Kempelen and the empress chatted, Kempelen drawing on his scientific knowledge to explain how the tricks worked. He was not at all impressed by what he saw. Indeed, he seems to have been rather irritated by the Frenchman's sneering and condescending tone, with its implication that it was the role of France to instruct the other nations of Europe in scientific matters. Once the show was over, Maria Theresa asked Kempelen, in his capacity as a scientific expert, his opinion of the performance. To the surprise of everyone present, Kempelen calmly responded that he believed himself capable of constructing a machine, the effect of which would be much more surprising, and the deception far more complete, than anything the empress had just witnessed.

Kempelen was known as a dependable and serious person, so this impetuous claim seemed entirely out of character and was greeted with laughter. But Kempelen was not joking. The empress could hardly allow such a boast to pass without comment, particularly since the matter was now one of national pride. Excusing him from his official duties in Banat and Vienna for six months, she challenged Kempelen to deliver on his promise and to build an automaton more impressive than anything that had been seen in any of the courts of Europe. Kempelen agreed not to return until he was ready to stage a performance of his own.

He went back to his home in Pressburg, where he lived with his second wife, Anna Maria, and their young daughter, Theresa. Abandoning his usual duties, he retreated to his workshop, where he spent the next few months fashioning wood, brass, and clockwork machinery into the chess-playing automaton that would unexpectedly ensure his place in history. By the end of the allotted six months, Kempelen was ready to transport his automaton to Vienna for its debut. By keeping his word, outdoing Pelletier, and impressing the empress, he could expect to be well rewarded. But things did not turn out quite as Kempelen imagined.

CHAPTER TWO

The Turk's Opening Move

THE QUEEN'S KNIGHT OPENING (E4 E5 NC3): An opening in which White advances his king's pawn and his queen's knight. Also known as the Vienna Game.

The most daring idea that a mechanician could ever conceive would be without doubt that of a machine which would imitate by more than mere form and movement the masterpiece of all creation. Not only has Mr. von Kempelen conceived such a project, he has executed it, and his chess-player is without any contradiction the most amazing automaton which has ever existed.

– Chrétien de Mechel, from the preface to Carl Gottlieb von Windisch, *Letters on Kempelen's Chess Player* (1783)

Wolfgang von Kempelen was a man of many talents. In addition to his linguistic, administrative, and technical skills, he proved to have an unusual gift as a showman. At the debut appearance of his chess-playing automaton before the Viennese court, Kempelen put on a theatrical performance that, with very little modification, would continue to amaze audiences for decades to come.

The automaton's first appearance took place in the spring of 1770 before a select gathering including the empress herself. The audience probably included many of the people who had witnessed the conjuring show six months earlier at which Kempelen had made his audacious claim, and who now expected to see him humiliated.

When Maria Theresa indicated that he should begin, Kempelen brought his automaton in from a side room and wheeled it forward for closer scrutiny. It consisted of a wooden cabinet, behind which was seated a life-size figure of a man, made of carved wood, wearing an ermine-trimmed robe, loose trousers, and a turban – the traditional costume of an oriental sorcerer. This choice of outfit was also highly fashionable at the time; the Turkish style, with its alluring combination of elegance and exoticism, was wildly popular in Vienna, where it was all the rage to drink Turkish coffee, dress one's servants in Turkish costumes, and appreciate the addition of supposedly Turkish touches (such as drums and cymbals) to contemporary music. As well as being stylish, the wooden figure's Turkish garb suggested knowledge of the game of chess, which had reached Europe from Persia sometime between 700 C.E. and 1000 C.E.

The cabinet itself measured four feet long, two and a half feet deep, and three feet high, and sat on four brass castors, one in each corner, which raised it off the floor slightly. This meant that it could be moved around and rotated freely, carrying the seated figure along with it, so that the whole contraption could be easily viewed from every angle. The front of the cabinet was divided into three doors of equal width, underneath which was a long drawer. The wooden figure had its right arm extended and resting on the top of the cabinet, and its eyes stared down fixedly at a large chessboard screwed on to the tabletop. In its left hand it held a long Turkish pipe, as though it had just finished smoking.

Stepping forward to address the audience, Kempelen announced that he had built a machine the likes of which had never been seen: an automaton chess player. A sceptical murmur passed through the audience. Kempelen explained that before demonstrating his automaton, he would display its inner workings. He reached into his pocket and produced a set of keys, one of which he used to unlock the leftmost door on the front of the cabinet. Kempelen opened this door to reveal an elaborate mechanism of densely packed wheels, cogs, levers, and clockwork machinery, prominent among which was a large horizontal cylinder with a complex configuration of protruding studs on its surface, similar to that found in a clockwork music box. As the audience scrutinized these workings, Kempelen walked around to the back of the cabinet. There he unlocked and opened another door directly behind the machinery. He then held a burning candle behind the cabinet in such a way that its flickering light was just visible to the audience through the intricate clockwork. Having checked that the onlookers had seen right through the cabinet, Kempelen then closed and locked the rear door.

Returning to the front of the automaton, Kempelen unlocked and opened the long drawer to reveal a set of chessmen, in red and white ivory, which he placed on the top of the cabinet. Next, he unlocked and opened the two remaining doors in the front of the cabinet to reveal the main compartment, comprising the rightmost two-thirds of the interior of the cabinet, which contained a red cushion, a small wooden casket, and a board marked with gold letters. Kempelen placed these items on a small table near the automaton, leaving the doors open to allow the audience to examine the main compartment closely. It was lined with

Carl Gottlieb von Windisch's engraving showing the Turk with its doors open.

dark cloth and was mostly empty, apart from a few metal wheels and cylinders, and two horizontal brass structures resembling quadrants.

Going back behind the automaton, Kempelen opened another door in the rear of the cabinet that opened into the main compartment. It was thus possible to see right through the main compartment, a point Kempelen demonstrated by holding a lighted candle behind the cabinet so that it could be seen by the audience. Still holding the

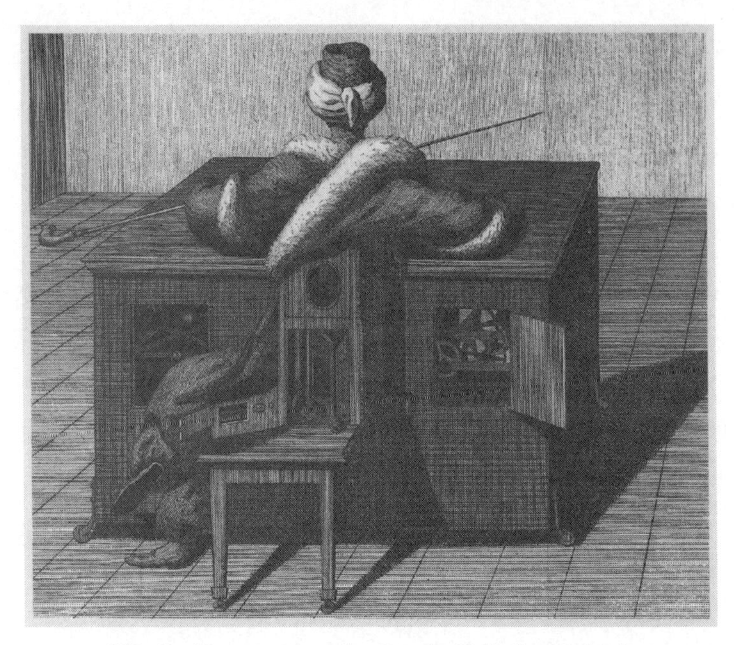

Windisch's engraving showing the Turk from behind.

candle, he then reached into the main compartment from behind, illuminating it in such a way that it was possible to see right into its darkest corners.

Leaving all the doors and the drawer open, Kempelen then rotated the automaton, causing the doors to flap on their hinges, so that the Turkish figure had its back to the audience. Lifting up the figure's robe and throwing it over its head, he revealed two more small doors in its left thigh and its back, both of which he opened to reveal more clockwork machinery. The automaton was then rotated and wheeled around so that everyone in the audience could have a good look at it from all sides. After this, Kempelen closed all the doors and the drawer, replaced the figure's robe, and returned the automaton to its original position facing the onlookers. He placed the cushion beneath the figure's left elbow, removed the long pipe from its left hand, placed the chessmen on the appropriate squares of the chessboard, and reached inside the cabinet to make a final adjustment to the machinery. Finally, he put two candelabra, each holding three burning candles, on top of the cabinet to illuminate the board.

Kempelen then announced that the automaton was ready to play chess against anyone prepared to challenge it, and recruited a volunteer – a courtier called Count Cobenzl – from the audience. Kempelen explained that the Turk would play with the white pieces and would have the first move, that moves could not be taken back once made, and that it was important to place the chessmen exactly on the centre of the squares when moving them. He explained that this was to ensure that the automaton would be able to grasp them correctly, without damaging its fingers. The count nodded. Kempelen then walked around to the left-hand side of the automaton, inserted a large key into an aperture in the cabinet, and wound up the clockwork mechanism with a loud ratcheting sound.

Once Kempelen had stopped turning the key, there was an agonizing silence. Then, after a brief pause, the sound of whirring and grinding clockwork, like that of a clock preparing to chime, could be heard coming from inside the automaton. The carved wooden figure, which had hitherto been completely immobile, slowly turned its head from side to side for a few seconds, as though surveying the board. To the utter astonishment of the audience, the mechanical Turk then suddenly lurched into life, reaching out with its left arm and moving one of its chessmen forward. The audience cried out in amazement. The game had begun.

*

As it made each move, the Turk's gloved left hand moved over the board so that it was positioned above a particular chessman. Its fingers would then close to grasp the chessman and move it to another square (or off the board entirely, when taking an adversary's piece). After making each move, the automaton rested its arm on the cushion, at which point the sound of whirring clockwork would cease.

As well as moving its left arm, the Turk moved its head under certain circumstances during the game. After making a move endangering its opponent's queen, the Turk would nod its head twice; and when placing the king in check, it would nod three times. In the event of its opponent making an incorrect move, such as moving a knight as though it were a bishop, the Turk would shake its head, move the offending piece back to its original square, and then proceed with its own move, thus forcing its opponent to miss a turn.

Windisch's engraving showing the Turk playing chess.

Every ten or twelve moves, Kempelen returned to the left-hand side of the cabinet to wind up the clockwork mechanism. Apart from this, he did not touch the Turk during the game but confined himself to standing nearby and occasionally peering into the wooden casket he had earlier removed from its interior, which had a hinged lid to prevent anyone from seeing its contents. Somewhat mischievously, Kempelen did not explain the function of the casket, so it was assumed by some onlookers to be a magical device by which he controlled the Turk's movements.

As if the sight of a machine playing chess was not astonishing enough, the Turk proved to be a formidable player. Count Cobenzl was swiftly defeated; the automaton was a fast, aggressive player and subsequently proved to be capable of beating most people within half an hour. Kempelen then further amused the audience by getting his automaton to solve some chess puzzles. In particular, the Turk was able to solve a classic problem called the Knight's Tour.

The Knight's Tour is one of the oldest puzzles in the history of chess; it has even been suggested that the tour is older than the game itself, and that the knight's move in chess is derived from the tour. It is difficult to solve but easy to state: the aim is to find a route that will carry a single knight (making its characteristic L-shaped move) around an empty chessboard so that it visits each square once, and only once. Try moving a knight around a chessboard, placing a marker (such as a dried pea) on every square it visits, and you will soon discover how difficult this is. To further complicate matters, it is sometimes specified that the knight should return to the starting square at the end of its tour. This is called a re-entrant tour.

Solving the Knight's Tour was of particular interest to mathematicians in the eighteenth century, several of whom derived solutions to it through elaborate analysis, rather than mere trial and error. But to be presented with a chessboard and told to demonstrate the Knight's Tour starting on a particular square was still more than most chess players could manage. Kempelen's mechanical Turk, however, could perform this complex and repetitive task with ease. A spectator chosen from the audience was asked to select any square on the board and place the knight on it. The automaton would then immediately reach out and take the knight, and move it around the board, visiting each square only once. To prove that this was so, the spectator would be asked to place a white counter on the starting square and a red counter on each square subsequently visited by the knight on its journey around the board.

*

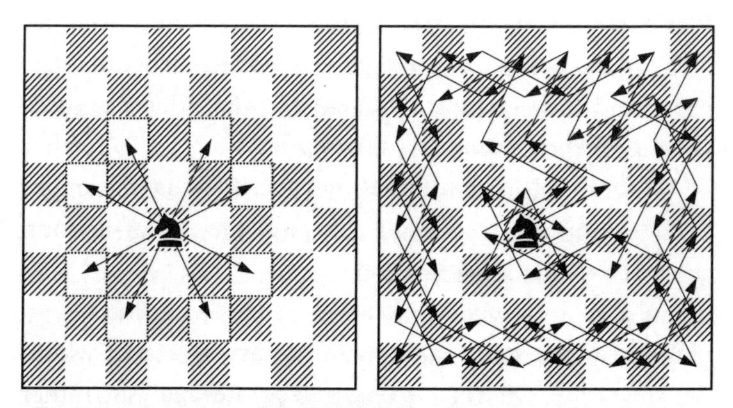

The knight's move, and the Knight's Tour as performed by the Turk. The knight visits every square once and only once before returning to its starting position. This pattern can be used to complete the tour from any starting position.

The Turk's sensational performance astonished and delighted the empress, and at her behest Kempelen and his automaton made many more appearances before additional members of the royal family, government ministers of both Austria-Hungary and foreign countries, and other eminent visitors to the court. Kempelen's extraordinary creation became the talk of Vienna, and it soon became more widely known as letters describing the automaton were published in newspapers and journals overseas.

One of the most widely circulated early descriptions of the Turk was that written by Louis Dutens, a traveller who saw a demonstration of the automaton at Kempelen's home in the summer of 1770, a few months after its debut in Vienna. On July 24 Dutens wrote a letter to the editor of a French newspaper called *Le Mercure de France.* This letter, describing the automaton, appeared in print in Paris that October and subsequently appeared in English in a London periodical, the *Gentleman's Magazine.*

Dutens explained that he had made Kempelen's acquaintance during a visit to Pressburg and had been greatly impressed by his engineering prowess. "It seems impossible to attain a more perfect knowledge of mechanics than this gentleman has done," he wrote. "No one has been able to produce so wonderful a machine as the one he constructed a year ago . . . an automaton which can play chess with the most skilful players."

After describing the appearance of the Turk, Dutens

explained Kempelen's willingness to reveal its insides, "especially when he finds any one suspects a boy to be in it. I have examined with attention all the parts both of the table and the figure, and I am well assured there is not the least ground for such an imputation." He went on to describe his own experience in playing against the automaton at a demonstration staged for the English ambassador, an Italian prince, and several English lords. He was, he wrote, particularly struck by the precision of the movements of the automaton's arm: "It raises this arm, it advances it towards that part of the chessboard, on which the piece stands which ought to be moved; and then by a movement of the wrist, it brings the hand down upon the piece, opens the hand, closes it upon the piece in order to grasp it, lifts it up, and places it upon the square it is to be removed to; this done it lays its arm down upon a cushion which is placed by the chessboard."

Like many of the Turk's adversaries, Dutens decided to see what would happen if he tested the machine by making a false move. "I attempted to practise a small deception, by giving the Queen the move of a Knight; but my mechanic opponent was not to be so imposed upon; he took up my Queen and replaced her in the square from which I had moved her." He also noted that some people did not take kindly to the suggestion that a mere machine might be able to beat them at chess: "I have met several people who played neither as quickly, nor as well as the Automaton, but would even so have been greatly affronted to have been compared to it."

Although he had seen the automaton close up on several occasions and had played against it himself, Dutens could not explain how it worked. "Notwithstanding the minute attention with which I have repeatedly observed it, I have not been able in the least degree to form any hypothesis which could satisfy myself," he wrote. Kempelen, he explained, moved around during the game, and at times was as much as five or six feet away from the automaton, so it was hard to imagine how he could have influenced its movements. Some people, he went on, "who had seen the effects produced by the lodestone in the curious exhibitions on the Boulevards at Paris", thought the Turk might rely on magnetism. But Dutens thought this was unlikely, because Kempelen, "with whom I have had long conversations since on this subject, offers to let anyone bring as close as he pleases to the table the strongest and best-armed magnet that can be found, or any weight of iron whatever, without the least fear that the movements of his machine will be affected or disturbed by it".

Dutens's letter to *Le Mercure de France* prompted an exchange of letters in its pages, as other correspondents wrote in to give their own views on the Turk and argue about how it worked. One young man, who had not seen the automaton himself, wrote in to argue that a machine could not make spontaneous movements, and that the automaton must therefore have had a child concealed inside it. Dutens angrily responded that he and several other learned gentlemen who had observed the Turk first hand had established that the interior not only was incapable of concealing a child but did not even have enough room for "a small monkey". He gave this assurance because a few months earlier there had been a report that the sultan of Baghdad had a chess-playing monkey, which some people had suggested might have since entered Kempelen's employ. In response to another letter, Dutens emphasized his belief that Kempelen himself was controlling the automaton. He admitted he had no idea how, but insisted that his own inspection of the automaton ruled out the possibility that Kempelen was directing the machine via magnetism or by pulling on tiny threads like a puppeteer controlling a puppet.

This exchange of letters highlighted a theme that was to reappear repeatedly throughout the Turk's career: the question of whether a machine could make unplanned movements under its own initiative. Most automata, after all, simply did the same thing over and over again; had the Turk done nothing more than mechanically execute the Knight's Tour, for example, it would have fallen into the same category. It was the automaton's apparent ability to respond to the moves of its human opponent during a game – its *interactivity,* to use the modern term – that set it apart from previous automata. Dutens concluded that "the wheels and springs make planned movements, but under the control of an unknown directing force."

*

The publication of Dutens's letters caused news of the automaton to spread throughout Europe, somewhat to Kempelen's embarrassment; he had not expected it to cause such a sensation. He had, however, achieved his objective. For the ingenuity of the Turk so impressed the empress that she recalled Kempelen from Banat and awarded him a generous additional allowance equal to his annual salary. More important, now that the Turk had proved Kempelen's engineering prowess, she gave him several engineering tasks to perform in addition to his court duties. In 1772 he constructed an elaborate hydraulic system to power the fountains at Schönbrunn palace, and two years later he built an elaborate mechanical bed (in effect, an elevator) for Maria Theresa, who was unwell. He also designed bridges, invented equipment for canal building, and experimented with steam engines. The success of the Turk launched Kempelen's parallel career as an engineer and man of science, which he pursued in addition to his duties as a civil servant.

In the years following, Kempelen did his best to distance himself from the Turk. As far as he was concerned, it had now served its purpose. He started to tell people that the automaton had been damaged in order to avoid having to demonstrate it, though he made occasional exceptions for important visitors. In August 1774 he wrote to Sir Robert Murray Keith, a Scottish nobleman who insisted on seeing the automaton, that it had gone wrong after being damaged in transit, but that he expected to be able to repair it within a couple of weeks. Later that month, Sir Robert explained in a letter to a friend, he and fifteen of his countrymen arrived at Kempelen's house at Pressburg "to see the famous automaton, which (as you may have read in every newspaper of Europe) plays at chess without the help of any visible agent, without any persons being concealed in or near it". The Turk played two games, winning the first but, unusually, losing the second.

Like most other observers, Sir Robert was unable to explain how the Turk worked. "There is no telling you how strange a thing this automaton is, nor how very perfect in all its operations. If the lodestone be the principle of its movements, it is at least so well concealed that there is no guessing its secret; and I am assured that magnets, and bars of iron, have been brought on purpose to counteract its motions, without having produced that effect." Evidently this was the Turk's first performance for some time, for Sir Robert noted that "the automaton had been dismounted for some years, because, I suppose, it took up too much of the gentleman's time, but he put it in order at my particular request."

Kempelen was clearly eager to put his reputation as the creator of the Turk behind him. His friend Carl Gottlieb von Windisch wrote that he "refused the entreaties of his friends, and a crowd of curious persons from all countries, the satisfaction of seeing this far-famed machine". Kempelen insisted that he was a civil servant and an engineer, not an entertainer. Proud of the fact that nobody had ever guessed its secret, and unwilling to divulge it, he also refused to sell the automaton to anyone else. Kempelen, according to Windisch, "thinking himself sufficiently repaid by the praises which this machine had acquired for him, and wishing to enjoy still longer the pleasure of alone possessing the secret, rejected several offers which were made to him of considerable sums, by persons who founded upon its acquisition various speculations of a pecuniary nature".

Sometime after Sir Robert Murray Keith's visit, Kempelen dismantled the automaton, and for several years he left it to gather dust. By the time of Maria Theresa's death in November 1780, it had been forgotten about altogether. Kempelen believed the Turk's career was over; but it had, in truth, only just begun.

CHAPTER THREE

A Most Charming Contraption

THE PARIS OPENING (NH3): An unusual opening in which White's first move places his king's knight at the edge of the board, rather than in a more central position.

The game of chess is not merely an idle amusement. Several very valuable qualities of the mind, useful in the course of human life, are to be acquired or strengthened by it, so as to become habits, ready on all occasions. For life is a kind of chess, in which we have often points to gain, and competitors or adversaries to contend with, and in which there is a cast variety of good and ill events, that are, in some degree, the effects of prudence or the want of it. By playing at chess, then, we may learn.

– Benjamin Franklin, *The Morals of Chess*

In 1781, more than a decade after its triumphant debut, the chess-playing Turk was languishing in its creator's home, neglected and dismembered. Wolfgang von Kempelen had turned his attention to several other projects, including a typewriter suitable for use by the blind and an ambitious attempt to build a machine capable of imitating the human voice. But the Turk was soon to be called back from retirement, due to the impending arrival in Vienna of Grand Duke Paul of Russia. Paul, the son of Catherine the Great and heir to her throne, had set off on a tour of Europe with his wife, Maria Fyodorovna, in September 1781, and Vienna was to be the first major stop on the journey. The emperor Joseph II, who had succeeded his mother, Maria Theresa, upon her death the previous year, was determined to do everything in his power to impress his visitors. What unique form of entertainment could he provide that would outshine anything the grand duke and duchess would see elsewhere? It was not long before Joseph remembered the chess-playing automaton and the excitement it had caused a few years earlier. He duly ordered

Kempelen to repair his automaton in time for the arrival of the party from Russia.

Kempelen had mixed feelings about rebuilding the Turk but had little choice except to comply with this request. Within five weeks he had completely refitted and reconstructed the automaton. Its appearances before the grand duke and duchess during their seven-week stay in Vienna were an unqualified success and prompted the grand duke to propose that Kempelen take his automaton on a tour of Europe – a suggestion that was greeted with enthusiasm by the whole court, though not by Kempelen. Joseph told him he would be excused from his duties in Vienna for two years for the purpose.

Having been reluctant to resurrect the Turk, Kempelen was even more averse to being pressed into service as a travelling showman. Initially he considered sending the automaton away in the care of a trusted associate, but he was concerned that if anything went wrong it would be necessary to rely on foreign craftsmen for repairs, and that they could not be depended upon to keep its secret. Besides, the offer of two years' leave, coming from the emperor, was effectively a direct order. The death of the empress had meant that Kempelen had lost his patron, not to mention his generous allowance. And, as his friends pointed out to him, his influence at the court had diminished, and his position was not improving. So Kempelen reluctantly agreed to the idea of a tour.

Altering the automaton for travel, so that it could be taken apart, packed, transported, and then put back together again, took several months. To Kempelen's chagrin, it also put a stop to his work improving his speaking machine. But by the beginning of 1783 the preparations were complete and the Turk was ready to go on the road. It must have seemed to Kempelen that the automaton had a life of its own. He had brought it into the world – yet now it was controlling his destiny.

*

Of all the cities of Europe, two were renowned for their enthusiasm for chess during the eighteenth century: Paris and London. Chess had been a popular pastime in coffeehouses in both cities since the beginning of the century and enjoyed a period of heightened popularity in the 1770s and 1780s, when it became extremely fashionable in high society. As the nearer of the two cities to Vienna, Paris was the logical place for the first stop on the Turk's tour of Europe.

As the French writer Denis Diderot put it in 1761, "Paris is the place in the world, and the Café de la Régence the place in Paris, where this game is played best." The Café de la Régence was a coffeehouse founded in the 1680s, and by the 1740s it had become the most prominent haunt of chess players in the city. Well-known intellectuals who were regulars at the café over the years included the philosophers Voltaire and Jean-Jacques Rousseau, the American statesman and scientist Benjamin Franklin, and even the young Napoleon Bonaparte. One of the strongest players to be found at the café was Legall de Kermeur, who according to one account was "a thin, pale old gentleman who sat in the same seat in the café, and wore the same green coat, for a number of years". But the greatest of all was Legall's pupil François-André Danican Philidor, arguably the most gifted chess player of his day, and without question the strongest player in Paris. (There was a strict pecking order among the chess players of France; Legall and Philidor were considered "first rank" players, five others were regarded as "second rank", and inferior players were classified as members of the third, fourth, and fifth ranks.)

With the Turk packed into boxes and loaded onto a carriage, Wolfgang von Kempelen and his small entourage (which included his wife, Anna Maria, his daughter, Theresa, his son, Carl, and his manservant, Anthon) rolled into Paris in April 1783. The city presented both an opportunity and a threat. For although his automaton had proved an impressive player when pitted against members of the court in Vienna, Kempelen was well aware that the chess players of Paris were of a far higher calibre. There was also a strong possibility that his automaton would be subjected to the scrutiny of the members of the Academy of Sciences, one of Europe's foremost scientific societies. Such scientific gentlemen were as well qualified as anyone in Europe to fathom the Turk's secret.

The Turk's arrival was recorded by Louis Petit de Bachaumont, a chronicler of the French court in Versailles, in his entry for April 17. "There was talk in the foreign papers of an automaton chess player, though no-one has heard of it for quite a while. It has just arrived in Paris," he wrote. "A Mr. Anthon has brought it here from Vienna, and it will be put on display to the public for the first time on Monday 21st." Rather than present the Turk himself, Kempelen had decided that Anthon should assist him in demonstrating the automaton, at least some of the time. After giving a brief description of the Turk, Bachaumont concluded credulously that "this spectacle seems far superior to other previously-known automata, such as the digesting duck, the flute-player, etc. It performs not just physical motions, but elevated intellectual functions."

Before going on public display in Paris, the automaton spent a few days entertaining the court at the palace at Versailles, just outside the city. (Kempelen would have had no difficulty obtaining an introduction to the court since the queen of France, the ill-fated Marie Antoinette, was Joseph II's sister.) The Turk proved so popular that it ended up staying longer than expected in Versailles. On April 25 Bachaumont noted that the enthusiasm for "Mr. Anthon and his automaton" was such that it had still not been allowed to depart for Paris. Bachaumont also reported that the duc de Bouillon had volunteered to play against the automaton and had beaten it. His victory was, said Bachaumont, not so much due to his skill as to "the complacency of his adversary, which demonstrated intelligence and some impressive moves". Despite this rare defeat, the Turk's evident ability meant it was being taken seriously by the chess players of Paris. Indeed, some people even suggested that a match should be arranged with the great Philidor himself.

The Turk eventually went on public display in Paris at the beginning of May; spectators were charged a small admission fee. For the first time, the Turk could be seen, and challenged, by members of the public. On May 6 it was defeated by a lawyer named Mr. Bernard, one of the five second-rank players in France. The match took place in the presence of a large and distinguished audience, including the marshal de Biron and the marquis de Ximenes, and afterwards Bernard was asked what he thought of the automaton's style of play. He acknowledged that the automaton was a resourceful player, possibly of the third or fourth rank. Then, after considering for a moment, Bernard, who was a portly, slow-moving man, singled out the marquis de Ximenes in the audience and declared, "The automaton has the same ability as the Marquis." According to Friedrich Melchoir von Grimm, a French author and chronicler of Parisian society who recounted the tale in a letter to his friend Diderot, Bernard's pronouncement then proceeded to spread throughout the whole city, to the utter mortification of the marquis.

Grimm was very impressed by the automaton. "Physics,

chemistry and mechanics have produced in our age more miracles than fanaticism and superstition could have believed possible in the ages of ignorance and barbarism," he declared. The Turk, he suggested, "is to the mind and the eyes what Mr. Vaucanson's flute-player is to the ear". Grimm noted that the Turk was capable of performing the Knight's Tour, something that he claimed even Philidor was unable to do. He also reported that a number of scientific experts had come to watch the Turk in order to try to discover how it worked. But, he wrote, "our greatest scientists and most able engineers have had no more luck than their Austrian counterparts in discovering the means by which the automaton's movements are directed." Even so, unlike Bachaumont, who took the automaton at face value, Grimm was convinced that it was being surreptitiously directed by a human operator. "The machine," he noted, "would not know how to execute so many different movements, which could not be determined in advance, unless it was under the continual control of an intelligent being."

*

During his stay in Paris, Kempelen was particularly eager to meet Benjamin Franklin, the American statesman and scientist. In 1783, Franklin had been a signatory of the Treaty of Paris, which marked the end of the American Revolution, and was the United States' formal representative in France. Before leaving Vienna, Kempelen had asked a friend named Valltravers to write him a letter of introduction. Valltravers wrote to Franklin that Kempelen was "endowed with peculiar taste and genius for mechanical inventions and improvements for which he sees no manner of encouragement in these parts" and explained that "he means to import several of his most important discoveries and experiments wherever they shall be best received and rewarded." Kempelen had, in other words, come to see the Turk's tour of Europe as an opportunity to meet other scientists, discuss his ideas and inventions, and promote himself within the scientific community.

Conveniently for Kempelen, Franklin happened to be a chess fanatic in addition to his many other interests. Indeed, such was his passion for the game that he sometimes played

Benjamin Franklin

for as many as five hours a day. On one occasion he played chess late into the night and sent out for fresh candles so that his game would not be interrupted, only to be told that dawn had long since broken; according to another story, Franklin once put off opening an important letter from America until he had finished his chess game. He also wrote a pamphlet on the subject, titled *The Morals of Chess.*

On May 28 Kempelen wrote to Franklin directly, inviting him to visit his rooms in the Hotel d'Aligre in order to examine both the chess player and his half-finished speaking machine, which he also had with him. Franklin visited him a few days later and played a game against the Turk, which defeated him. No record of the encounter can be found among Franklin's papers, however, though this may be partly explained by the fact that Franklin is known to have been a sore loser. But there is indirect evidence of his encounter with the Turk: a letter Franklin received from a friend in October 1783 thanked him for a previous letter that had been delivered in person by Kempelen. And many years later his grandson related that Franklin had greatly enjoyed his game against the automaton.

The Turk's public performances in Paris continued throughout the summer. On June 24 the duc de Croy, a French nobleman, recorded in his diary that "finding myself at a loose end, I went to see the automaton chess player. It was well worth the trouble. It is a most charming contraption, completely freestanding, that works by a trick that no one has been able to discover, which is most curious." The duke may have seen the Turk at the Café de la Régence, where the automaton was said to have played (and lost) a match against Legall. Win or lose, however, the Turk continued to draw in the crowds. And, after weeks of anticipation, a match was finally arranged with Philidor.

*

Philidor had only just returned to Paris from London, where he spent the period from February to May each year playing at a chess club called Parsloe's. It was in London in May 1782 that Philidor gave the first of many exhibitions in which he played two opponents simultaneously while blindfolded, something regarded as an extraordinary achievement at the time. On May 28, 1783, he took on three opponents at once while blindfolded, beating two and ending in a draw with the third. One London newspaper described this as "a wonder of such magnitude, as could not be credited, without repeated experience of the fact".

Even though he was generally recognized in England, France, and elsewhere as the best chess player in Europe, Philidor primarily regarded himself as a musician. Born in 1726, he came from a musical family, sang in the choir of the chapel royal in Versailles as a boy, and learned chess at the age of ten from one of the court musicians before going on to study with Legall. (The court musicians were forbidden from playing games of chance, such as gambling with cards or dice, but were allowed to play chess; they had a long table for the purpose, inlaid with six chessboards.) It was in 1745, while on a musical tour in Rotterdam that was cancelled when the soloist fell ill, that Philidor had to fall back on his chess playing as a means of earning a living. He was a talented player, but what really made his name was the publication of his book, *The Analysis of Chess*, a detailed and comprehensive guide to chess strategy published in 1749 that was instantly recognized as a classic.

Philidor also pursued a parallel career as a composer. As a boy, one of his compositions had been performed in the presence of Louis XV, and the quality of his subsequent work was acknowledged by Handel. His best-received works were his comic operas, but none of them made any money, and he had to subsidize his musical career by teaching chess and playing exhibition matches, including blindfolded games.

In person, Philidor was a bumbling, amiable, though somewhat humourless character; it was once said of him that he had "no common sense – it is all genius". While playing chess he would fidget constantly, moving his legs under the table. His style of play was, however, extremely innovative, and several chess tactics are named after him. In particular, he emphasized the importance of the pawns and was fond of making strategic sacrifices; the term *Philidor Sacrifice* is used to describe the sacrifice of a knight or bishop in order to improve the position of one's pawns. Another daring tactic is known as Philidor's Legacy in his honour. It involves the sacrifice of the queen, the most powerful piece, in an audacious combination attack that leads to checkmate and thus victory at the next move.

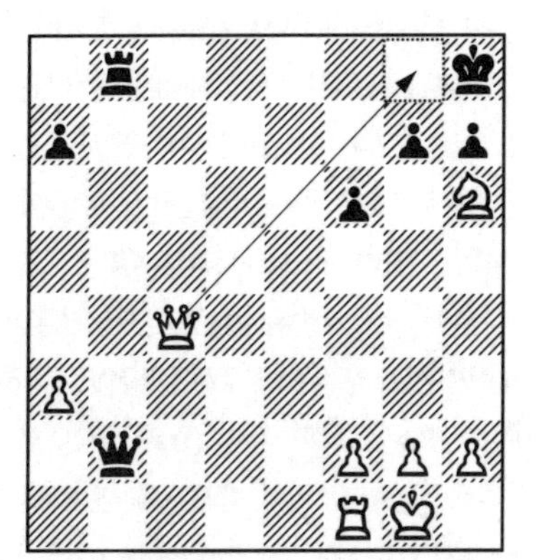

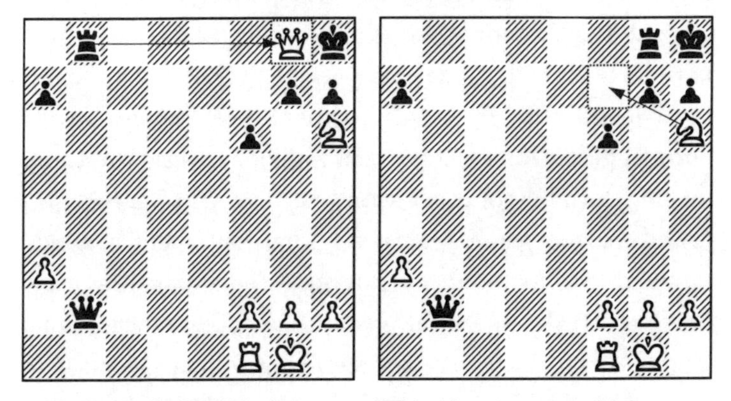

Checkmate by Philidor's Legacy. White, in a strong position, uses his queen to attack the black king. Black responds by taking the white queen with his rook. This hems in the black king, which White then checkmates with his knight (a "smothered mate"). The sacrifice of the white queen thus leads to victory.

The very fact that Philidor agreed to play against the Turk was a coup for Kempelen. But he knew that a victory over the master would be even more valuable. According to the somewhat implausible account of Philidor's eldest son, André, Kempelen approached Philidor the day before the match was due to be played with an unusual proposal. "I am not a magician, Sir, you know that, and my automaton is not stronger than I," he is supposed to have said. "It is, moreover, my only means of earning a living at present. Imagine, then, what it would be worth to me to be able to announce and publish in the newspapers that my automaton had defeated you." Philidor was not a vain man and apparently agreed to allow the machine to beat him if it played well enough for such a victory to be plausible. But, he said, if it did not prove to be a strong player, he would have no qualms about defeating it.

Evidently the Turk was nowhere near being a strong enough player for a victory over Philidor to have been convincing, for Philidor easily won the game. But he later confessed that no game against a human opponent had ever fatigued him to the same extent. Philidor apparently believed the Turk was entirely genuine, and found the idea of a chess-playing machine rather terrifying. Perhaps his attitude – and that of others in Paris who regarded the Turk as a true automaton – was not surprising, given the intellectual climate in the summer of 1783. News of the first public demonstration of a hot-air balloon, given by the Montgolfier brothers at Annonay in southern France on June 5, had caused a sensation in Paris. If flying machines could be built, why not a thinking, chess-playing machine?

Philidor's match with the Turk took place in the presence of several members of the Academy of Sciences, who were still doing their best to discover the automaton's secret. According to an account that appeared in the September edition of the *Journal des Savants,* a Parisian publication that has the distinction of being the world's first modern scientific journal, they did not succeed. The Turk, the report explained, "is described quite overtly in certain newspapers as though it were an automaton that really plays, and on its own; what is certain is that the manner in which its maker influences his machine during play is so adroit and so well hidden that a large number of savants who saw it at Paris were not able to divine the means by which it is done."

The best the Parisian experts could do was to form a detailed hypothesis about how the Turk might work. Their explanation was certainly more convincing than vague insinuations that magnets, or hidden children, were responsible for its operation. The experts concluded that the Turk relied on a combination of magnetic trickery and mechanical ingenuity. Noting that Anthon returned to the side of the machine from time to time to wind it up, and having observed the elaborate studded cylinder inside the cabinet, they suggested that the cylinder encoded sets of preprogrammed chess moves, just as similar cylinders seen inside musical automata (such as Vaucanson's flute player) encoded the tunes they could perform. Kempelen was assumed to use a magnet hidden in his pocket to switch between these sets of moves. In other words, he did not control the automaton on a move-by-move basis but merely steered its strategy; this would explain why he could leave the automaton to play unattended for several moves at a time. All of this was, noted the *Journal des Savants,* "only a vague impression, which can only increase the admiration due to the truly extraordinary talents of Mr. Kempelen".

For Kempelen, the match with Philidor had been a public-relations triumph, despite the Turk's defeat. Riding high on the honour of having taken on the finest player in Europe, and with the secret of his automaton still intact despite the high-powered scrutiny of the Academy of Sciences, he swept out of Paris and headed for London.

CHAPTER FOUR

Ingenious Devices, Invisible Powers

THE ENGLISH OPENING (C4): An opening move in which White advances his queen's bishop's pawn. In modern times, it is the third most popular opening move in world championship games.

Chess is the gymnasium of the mind.

– Peter Pratt, "Studies of Chess", 1803

It is difficult to imagine an attraction more likely to appeal to the Londoners of 1783 than Kempelen's chess-playing automaton. For in addition to being a great centre for chess, London was renowned for its enthusiasm for public displays of automata and other technological marvels. The arcades of Piccadilly, the streets of St. James's, and the squares of Mayfair were home to several remarkable exhibitions of automata and other curiosities, open daily to the paying public.

By the 1780s the best-known venue for such displays was a disused chapel in the Spring Gardens, near what is now Trafalgar Square. Its original proprietor was James Cox, the automaton maker who supplied expensive baubles (including jewelled elephants, elaborate clocks, and mechanical tigers, peacocks, and swans) to the East India Company, which sent them to China in exchange for tea. Unfortunately for Cox, the Chinese market eventually became so glutted that demand for his wares fell sharply, and in 1772 he was forced to sell off some of his stock to raise money. In the same year he also opened a museum as a means of deriving an income from his remaining automata. Cox spared no expense in decorating his museum; it had an elaborate painted ceiling, five crystal chandeliers, and crimson curtains, and was filled with his characteristic jewel-encrusted creations. The English writer Samuel Johnson, who visited the museum in April 1772, remarked that "for power of mechanism and splendour of show, it was a very fine exhibition."

Despite the exorbitant admission fee of ten shillings and sixpence, Cox's Museum was the talk of London during the three years it remained open. The same building subsequently became the home of Davies's Grand Museum, which was set up by one of Cox's assistants, a man named Davies. According to a newspaper account of May 1782, the exhibits included a silver-plumed swan, a boy "with a Pine-Apple on his Head, which opens and discovers a nest of birds", and a mechanical star decorated with 3,000 precious stones.

Another of Cox's protégés, a Belgian maker of automata named John Joseph Merlin, also established his own exhibition. For two shillings and sixpence, visitors to his Mechanical Museum in Hanover Square could see a gambling machine, several supposed perpetual-motion clocks, a number of music boxes, and even a therapeutic chair of Merlin's own design that was said to be able to help sufferers from gout. Merlin was a prolific inventor. His creations included an early form of roller skates, a combination piano-harpsichord, and an extraordinary rotating tea table: by pressing a pedal, the hostess could cause each of eight cups on the table to appear in front of her in sequence; a second pedal opened and closed the spout of a tea urn.

Such inventions were undoubtedly ingenious, even if they were not terribly practical. But Merlin and other automaton makers were well aware that the new technology could be used for far more than just entertainment. In addition to his therapeutic chair, for example, Merlin devised several other inventions intended to make life easier for elderly, sick, or disabled people. Another consequence of the rivalry between automaton makers was the development of new machine tools (such as the milling machine) and novel techniques (such as the use of multiple cams on a single shaft to enable complex, synchronized movements) that could also be applied in industry.

An etching by Thomas Lane of an automaton exhibition in the Gothic Hall, London.

As the industrial revolution gathered pace in the last quarter of the eighteenth century, many things that had previously been infeasible were suddenly possible. Balloons were taking to the skies in France and, soon afterwards, in other countries across Europe; in England, thanks to the innovations of James Watt and others, steam engines were no longer merely experimental devices but were increasingly practical for industrial use. Watt's contribution was to devise a steam engine in which the back-and-forth motion of the piston was converted into a continuous rotary motion. As a result, steam power, which had previously been used primarily for pumping, could be used to drive all kinds of other machinery whose rotating spindles and wheels would otherwise have been powered by waterwheels or the exertion of humans or animals. Furthermore, through a system of gears and rotating shafts, rotary power from a single steam engine could be distributed throughout a factory or workshop. Watt's steam engine thus provided the power source for increasingly complex mechanical devices and paved the way for industrial mechanization on a grand scale.

In such circumstances, as Johnson noted, the kind of displays of mechanical ingenuity that could be found in London were "unquestioningly useful, even when the things themselves are of small importance, because it is always advantageous to know how far the human powers have proceeded, and how much experience has been found to be within the reach of diligence . . . it may sometimes happen that the greatest efforts of ingenuity have been exerted in trifles, yet the same principles and expedients may be applied to more valuable purposes, and the movements which put into action machines of no use but to raise the wonder of ignorance, may be employed to drain fens, or manufacture metals, to assist the architect, or preserve the sailor."

This sentiment was later echoed in the 1830s by David Brewster, a British physicist and popularizer of science, who is also remembered as the inventor of the kaleidoscope. By the early nineteenth century the extent to which the previous generation's enthusiasm for automata had inspired advances in industrial machinery had become fully apparent. "The passion for automatic exhibitions which characterised the eighteenth century gave rise to the most ingenious mechanical devices, and introduced among the higher order of artists habits of nice and neat execution in the formation of the most delicate pieces of machinery," wrote Brewster. "Those wheels and pinions, which almost eluded our senses by their minuteness, reappeared in the stupendous mechanism of our spinning-machines and our steam-engines. Those mechanical wonders which in one century enriched only the conjuror who used them, contributed in another to augment the wealth of the nation; and those automatic toys which once amused the vulgar, are now employed in extending the power and promoting the civilization of our species."

At the intersection between entertainment, technology, and commerce, automata allowed new ideas to flow from one field to another and acted as a catalyst for further innovation. And during its visit to England, the Turk was to play an unexpected role in this process.

*

The Turk arrived in London in the autumn of 1783 and went on display at No. 8, Savile Row, Burlington Gardens. Its arrival in the city coincided with the publication of a promotional book written by Kempelen's friend and compatriot Carl Gottlieb von Windisch. In 1773 Windisch had published a brief description of the Turk, and in 1780 he described it more fully in a book about the Kingdom of Hungary. His 1783 book, initially published in German in the city of Basel, took the form of a series of letters to an anonymous friend, in which Windisch described the automaton, told the story of its origin, and gave some biographical details about Kempelen. Titled *Letters on Kempelen's Chess Player,* the book was also published in French in Paris, and it appeared in English soon afterward under the provocative title *Inanimate Reason.*

The letters in the book are dated from Pressburg between the seventh and thirtieth of September 1783, but since the book had already appeared in print in Paris by this time, these dates are clearly bogus. Indeed, the whole book was nothing more than a cleverly designed promotional device to encourage its readers to go in person to see the Turk, which was referred to as "indisputably the most astonishing automaton that has ever existed". Windisch described himself as an intimate friend of Kempelen, whose "astonishing genius" he went on to extol at great length. He described a performance by the automaton, noted the expressions of "extreme surprise" on the faces of the spectators, and made much of the inability of experts from all nations to discover the secret of the Turk's mechanism. Detailed engravings of the automaton, based on Kempelen's own drawings, were also included. Windisch ended his final letter with a blatant tease: "I can anticipate the eager desire you will feel to see for yourself all which I have just detailed concerning this machine."

Unlike the many other eyewitness descriptions of the automaton, Windisch's book is not to be trusted as an impartial account – it has Kempelen's fingerprints all over it – but it is valuable nonetheless as the nearest thing there is to a description of the automaton by its inventor. And it seems to have had the desired effect: the people of London were soon flocking to see the Turk, at a cost of five shillings each.

A review of the English translation of Windisch's book appeared in a London magazine, the *Monthly Review*. The author of the review poured scorn on those people "simple enough to affirm, both in conversation and in print, that the little wooden man played *really* and *by himself*". Notably, however, the chess-playing Turk was described as having "actually encountered and beaten the best players at that game; particularly, as we are informed, the celebrated Mr. Philidor". How the rumour got started that the automaton had beaten Philidor is unknown, but it certainly provided valuable publicity.

The anonymous author of the *Monthly Review* article was not the only person to lament the foolishness of those who regarded the Turk as genuine. Philip Thicknesse, a wealthy and well-travelled Englishman, was so incensed by Kempelen's display (which he attended with his family) that he went to the trouble of writing and publishing a pamphlet condemning it. Unlike the author of the *Monthly Review* article, he proposed an explanation for how the automaton worked.

Thicknesse was absolutely certain that the Turk was a trick, because it was able to respond to its opponent's moves – something that only a human could possibly do. "That an Automaton may be made to move its hand, its head, and its eyes, in certain and regular motions, is past all doubt; but that an AUTOMATON can be made to move the Chessmen properly, as a pugnacious player, in consequence of the preceding move of a stranger, who undertakes to play against it, is UTTERLY IMPOSSIBLE," he declared. "And, therefore, to call it an Automaton (a self-moving engine, with the principle of motion within itself) is an imposition, and merits public detection; especially as the high price of five shillings for each person's admission induces the visitor to believe that its movements are REALLY performed by mechanic powers." Kempelen's machine, said

Thicknesse, "when he is stripped of his Turkish robes, turned out of his splendid apartment, deprived of the serious deportment of all the parties, and parade of admittance, is a simple trick".

Before explaining how he thought the Turk worked, Thicknesse warned his readers not to be taken in by Kempelen's diversionary tactics. The display of the Turk's clockwork innards, he declared, was just one of many "ingenious devices, to misguide and to delude the observers". In particular, said Thicknesse, Kempelen set out to fool his audience into believing that he was secretly controlling the automaton "by some incomprehensible and invisible powers, according to the preceding move of the stranger who plays against the Automaton; and that every spectator should think so, he places himself close to the right elbow of the Automaton, previous to its move; then puts his left hand into his coat pocket, and by an awkward kind of motion, induces most people to believe that he has a magnet concealed in his pocket, by which he can direct the movement of the Turk's arm at pleasure. Add to this, that he has a little cabinet on a side-table, which he now and then unlocks, and locks; a candle burning; and a key to wind up the Automaton; all of which are merely to puzzle the spectators."

In fact, Thicknesse claimed, the Turk was not being controlled by Kempelen at all but by an operator concealed within the chest. "The real mover is concealed in the Counter, which is quite large enough (exclusive of the clockwork) to contain a child of ten, twelve, or fourteen years of age; and I have children who could play well at chess, at those ages," he wrote. The operator could see what was going on, Thicknesse suggested, by using a mirror attached to the ceiling; looking at the reflected image would make it possible for him to move the arm appropriately. To support this hypothesis, Thicknesse pointed out that the Turk was displayed only between 1 P.M. and 2 P.M., "because the invisible player cannot bear a longer confinement". Surely, he argued, Kempelen would have displayed his automaton for longer each day had it had been possible to do so, and made more money.

Thicknesse then rather undermined his argument by changing his mind as to the means by which the concealed operator could see the chessboard. "I saw the ermine trimmings of the Turk's outer garment move once or twice, when the Figure should have been quite motionless. I rather think the invisible player sees all the moves through the hair trimmings of the Turk's habit," he wrote.

This was not the first time that Thicknesse had set out to expose a mechanical deception that depended on a human operator concealed within a machine. "Forty years since, I found three hundred people assembled, to see, at a shilling each, a coach which went without horses; and thought this coach was moved by a man within," he recounted. "Many persons present were angry with me, for saying it was trod round by a man within the hoop, or hinder wheels; but a small paper of snuff, put into the wheel, soon convinced every person present, that it could not only move, but sneeze too, perfectly like a Christian. That machine was not a wheel within a wheel, but a Man within a wheel: and the Automaton Chess-Player is a man within a man; for whatever his outward form be composed of, he bears a living soul within." He concluded ruefully: "Man the cunningest, the most artful, and the most ingenious of all animals, is always aiming to deceive." Short of slipping some snuff inside Kempelen's automaton, however, there was no way for Thicknesse to prove that his explanation was correct, and his pamphlet seems to have done little to dent the Turk's popularity during the months it spent in London.

Another sceptical account of the Turk that appeared at around the same time was included in a book, *White Magic Exposed,* which was written by the Frenchman Henri Decremps and published in several languages across Europe. Decremps's book explained the workings of several popular conjuring tricks of the time, many of which exploited scientific principles that were unfamiliar to the general public. Most of the explanations consisted of a description of the presentation of a particular trick, followed by a detailed exposé of how it was done. This made for compelling reading, and Decremps's book was extremely popular. In the case of the Turk, however, Decremps (who had seen the Turk in Paris in 1783) could only guess at an explanation. Rather than claim to have discovered its secret, he presented his hypothesis in the form of a fictionalized exposé of a chess-playing automaton built by an inventor named Van Estin (whose name is clearly an allusion to von Kempelen).

Decremps concluded that there was a dwarf hidden inside the chest who operated the automaton. Initially, he suggested, the operator was hidden outside the chest, under the Turkish figure's robes; the noise made by winding the machine up was merely a cover to enable the dwarf to slip inside the chest once its empty interior had been shown to the audience. The automaton could then be spun around and the robes lifted. Once inside, Decremps theorized, the operator followed the game by looking up through the chessboard, which was semitransparent, or possibly via hand signals or verbal cues from Kempelen. The automaton's moves were then made by guiding its arm using a system of levers.

The main problem with Decremps's theory is that the automaton was always wound up after the rear of the chest had been shown to the audience and the figure's robes lifted. Furthermore, none of the players who examined the Turk up close noticed anything unusual about the chessboard, while the idea that Kempelen was constantly signalling to an operator inside the automaton did not stand up to close scrutiny, since Kempelen would often leave his automaton to play several moves while having a private conversation with a member of the audience.

Despite the scepticism of Thicknesse and Decremps, there were still many people who believed the Turk was entirely genuine. Charles Hutton, editor of the *Philosophical and Mathematical Dictionary,* described the Turk as "truly the greatest masterpiece of mechanics that ever appeared in the world".

*

In contrast with the sceptics who attempted to unmask the Turk as a fraud, some observers embraced Kempelen's automaton as an example of the unlimited potential of mechanical automation. Among them was Edmund Cartwright, a clergyman who enjoyed performing experiments and devising new inventions. He was the rector of a country parish in Leicestershire, where he carried out agricultural experiments on his land, including the development and testing of new agricultural implements of his own design. One summer, while on holiday, he visited the water-driven mills set up by Richard Arkwright, inventor of the spinning frame. Arkwright had patented the spinning frame in 1769, the same year that Kempelen built his automaton, and it went on to transform the textile industry. Arkwright's spinning frames improved the speed and efficiency of hand spinning, producing fabric that was firmer and smoother than that woven from traditional hand-spun cotton. By 1775 Arkwright and his associates had factories all over England, and by 1782 he was employing 5,000 workers. He was eventually rewarded for his contribution to industry with a knighthood from King George III.

In 1784, however, there was much concern surrounding Arkwright's patents, which were about to expire. This would mean that his spinning technology would become far more widely used, so that the production of cotton would far outstrip the capacity to weave it. Many inventors were, accordingly, working on machinery to automate the weaving process as well as the spinning process. But building a machine able to replicate the numerous complicated movements made by the hands and feet of the operator of a hand loom was extremely difficult. Indeed, even the great Vaucanson himself had attempted to promote the adoption of automatic looms, but without success.

Cartwright was unaware of this. In the summer of 1784 he was discussing Arkwright's spinning machinery with some businessmen from Manchester, one of whom stated his firm belief that weaving could never be automated. But Cartwright had just seen the Turk in London. Surely, he reasoned, if it was possible to construct a machine capable of playing chess, it ought to be possible to build an automatic loom. "Now, you will not assert, gentlemen," he replied, "that it is more difficult to construct a machine that shall weave than one that shall make all the variety of moves which are required in that complicated game." He set to work, and three years later he patented the power loom from which much subsequent weaving machinery was derived. The combination of the power loom and the steam engine made weaving possible on an industrial scale: in 1799 a factory was constructed in Manchester containing 400 of Cartwright's looms powered by steam. And, as the result of a petition organized by fifty prominent Manchester firms, Cartwright was later awarded the sum of £10,000 by Parliament in recognition of his invention.

Kempelen was unaware of his automaton's unwitting progeny. Having amazed London, enraged Thicknesse, and planted the seed for the power loom in Cartwright's mind, he and his automaton left England in the autumn of 1784 and continued their tour of Europe.

CHAPTER FIVE

Dreams of Speech and Reason

THE LEIPZIG VARIATION (D4 NF6 C4 E5 D×E5 NE4): An aggressive countergambit in which Black offers his king's pawn and then advances his king's knight beyond the white pawn that advances to take it.

There are more adventures on a chessboard than on all the seas of the world. – Pierre MacOrlan

Now homeward bound after a year and a half on the road, the Turk passed through Germany on its way back to Vienna, stopping in Carlsruhe, Frankfurt, and Gotha and arriving in Leipzig in September 1784. By this time it was at the height of its fame, thanks to both the popularity of Windisch's book (which had appeared in numerous editions throughout Europe) and the fact that the mystery of its operation remained unsolved. As a result, it was in Germany that the Turk came under the most intense scrutiny. It was seen at the Michaelmas Fair in Leipzig by Carl Friedrich Hindenburg and Johann Jacob Ebert, two scientifically minded men who independently went on to publish their own accounts of the automaton.

Both men noted that the Turk was presented by Kempelen and his assistant, Anthon. By this time Kempelen had become sufficiently confident in his assistant's ability to conduct the automaton's performances that he occasionally let him do so alone, while Kempelen himself sat in the audience. This further baffled those who supposed Kempelen to be directing the machine from a distance. In any case, the Turk's appearance at Leipzig was a great success. Ebert, who was professor of mathematics at the University of Wittenberg, declared in his account of the automaton that it and the speaking machine, which Kempelen also exhibited, were "the two most remarkable curiosities of the last Michaelmas Fair".

Ebert observed that the Turk was a strong player and lost only two games while in Leipzig. He discounted the possibility that there could be an operator concealed inside it, and since it was never next to a wall or a curtain during performances, he could see no way in which its movements could be governed by anyone except Kempelen or Anthon. Ebert concluded that the Turk's mechanical innards carried out its moves, but that its strategy was steered by its human operator. But how this guidance was carried out he could not explain; he thought magnetism was unlikely, because of Kempelen's willingness to allow a large magnet to be placed on top of the machine.

Hindenburg's report, "About the Chess Player of Mr. von Kempelen", appeared as a magazine article and then as a book within a few weeks of the fair. Little is known about Hindenburg himself, but his logical approach and the fact that his account was dedicated to Johann Bernoulli, a member of the scientific Bernoulli family, suggest that he (like Ebert) may have been an academic.

Hindenburg saw the Turk on two separate occasions and

came to essentially the same conclusion as the experts of Paris: rather than opt for an entirely magnetic or mechanical explanation for its operation, he suggested a combination of both. "An important part of the mechanism is undoubtedly the horizontal cylinder in the lower part of the chest, which by means of the studs on its surface, sets the adjacent lever in motion as it turns," he declared. This mechanical apparatus, he concluded, performed the actual moves, but under the guidance of the operator, who directed its strategy using some kind of magnetic trickery.

By this time Kempelen had adopted a new way of presenting his automaton. In a small space such as Kempelen's study, or when in front of a small crowd, the Turk encountered its human opponents across the chessboard, face-to-face. But this approach was not well suited to large public exhibitions of the automaton, because the human opponent tended to obscure the audience's view. So on such occasions Kempelen set up a second chessboard a short distance away from the automaton on a small table. The Turk's opponent would sit at this table and make his move; Kempelen would then make the equivalent move on the Turk's own chessboard. Once the Turk had responded with its own move, Kempelen would move for it on the second chessboard. This arrangement ensured that everyone could see the Turk clearly as it made its moves. It also enabled Kempelen to ensure that the chessmen were always placed in the centre of the squares, thus ensuring that the automaton would be able to grasp them correctly, without damaging its fingers.

Its secret still intact, the Turk moved on through Germany, visiting Dresden and a number of other towns. One of the many people who saw it in Germany was Johann Lorenz Böckmann, who came up with an entirely new theory about how it worked. Like Thicknesse, Böckmann assumed that there had to be an operator hidden inside the automaton itself. But this hypothesis raised a number of questions. Where was the operator concealed while the doors of the automaton were open? How would such an operator follow the action on the chessboard? And how could the operator direct the movements of the Turk's arm? Böckmann was interested only in the second of these three questions, and he proposed an ingenious mechanism by which it would be possible for someone hidden inside the Turk to figure out what was happening on the chessboard above. His novel idea was published as a short article, "A Hypothetical Explanation of the Famous Chess-player of Mr. von Kempelen", which appeared in *Magazin für Aufklärung* (literally, "The magazine for clearing things up").

"There is in my judgement under each square of the chessboard, a magnetic pointer suspended just off-centre, which is pointed in a given direction," he wrote. "Furthermore, each chessman contains a well magnetized piece of iron. Since magnetic force will penetrate all materials except iron, a chessman so armed would, when set down on a square, quite naturally change the direction of the pointer underneath, and set it into a noticeable movement. When a piece is lifted off its square, the pointer moves again and returns to its natural position. It follows therefore that a person sitting under the chessboard could easily and clearly observe the progress of the game." By the time Böckmann's article appeared in print in early 1785, the Turk had moved on to Amsterdam. Its tour then came to an end, and Kempelen and his automaton returned to Vienna.

*

After two years on the road, having acted as a technological ambassador for his country and discharged his duty to the emperor, Kempelen returned to his career at the imperial court. The Turk, packed into its wooden crates, was stored away for what would prove to be a lengthy slumber. It has been suggested that Kempelen wanted to retire his automaton before its secret was exposed by one of the innumerable pamphlets, books, or articles it seemed to inspire wherever it went. But it seems more likely that he had simply grown tired of the Turk and put it aside in order to pursue his other interests. Foremost among them was his research into the mechanism of speech, including the refinement of his speaking machine.

Kempelen constructed his first speaking machines in the 1770s after many years of research on the subject. He started off by building a device consisting of a tube, a reed, and a funnel; by placing his hand in different positions inside the funnel, he was able to produce vowel sounds. Kempelen's next machine was based on a detailed examination of the human organs of speech. Like his hero Vaucanson, Kempelen assumed that the best way to replicate biological functions artificially was to imitate the natural mechanisms as closely as possible. This new machine consisted of a hollow oval box made up of two overlapping sections joined by a hinge, so that they resembled a set of jaws. Connected to the box was a set of bellows and a reed. By opening and closing the jaws, Kempelen was able to produce the vowel sounds corresponding to the letters *A, O,* and *U,* though the *E* and *I* sounds were less convincing.

This machine was subsequently refined with the addition of a device that imitated the human glottis, so that it could produce the consonants *P, M, N,* and *L.* A set of levers controlled additional *S, Sh,* and *Z* sounds, and another lever dropped a wire onto the machine's reed to produce a rolling *R* sound. The consonants *D, G, K,* and *T* could be crudely imitated with variations of the letter *P.* Careful manipulation of the machine's controls could cause it to pronounce the whole words such as *mama, papa,* and *opera.* It was, in fact, the first machine ever built that was capable of saying entire words and even short, carefully chosen sentences. (Other speaking machines, such as that constructed by the Danish physiologist Christian Gottlieb Kratzenstein, which won a prize offered in 1779 by the Imperial Academy of St.

Petersburg, could pronounce only vowels.) Kempelen later added a pitch control as well, to allow the machine's intonation to be controlled; it had previously spoken in a droning monotone.

Operating the machine was akin to playing a musical instrument, since it required coordinated movement of the various controls. Indeed, Kempelen even designed a version of his machine with a conventional piano keyboard. But mastering the speaking machine was easier than learning to play a musical instrument; Kempelen estimated that a novice operator would take about three weeks to learn how to work it convincingly. He usually made the machine speak French, Italian, or Latin, all of which were far easier for the machine to imitate than German, which involves complicated combinations of consonants.

Ever the showman, Kempelen liked to amuse himself by concealing the machine on a table beneath a cloth when demonstrating it. Once a sentence had been proposed for the machine to speak, he would then reach underneath and operate the controls accordingly. Only then would he pull away the cloth to reveal the machinery beneath.

In his book of letters promoting Kempelen's chess automaton, Windisch also described the speaking machine. Windisch told the story of a young woman who entered a room and was greeted by Kempelen, who moved his lips but spoke using the machine instead of his own voice. The woman "was seized with such terror, as to be on the point of running away with the utmost speed; and it was not without the greatest of difficulty that he succeeded in reassuring her, by explaining whence the voice came, and enabling her to convince herself of the fact, by showing her the machine." This is, however, exactly the kind of story Windisch might have invented, since it perfectly suits his purpose of promoting Kempelen as the Hungarian equivalent of Leonardo da Vinci. Perhaps a more reliable firsthand witness of the speaking machine was the German poet and philosopher Johann Goethe, who commented that "the speaking machine of Kempelen is not very loquacious, but it can pronounce certain childish words nicely."

Kempelen ultimately published the results of his research into speaking machines in 1791, in a masterly book. (The English translation of its title is "The mechanism of human speech, with a description of a speaking machine".) As well as outlining his research, Kempelen's book included speculations about the nature of animal communications, the origins of language, and the relationship between the powers of speech and reason. He went on to describe the physical basis of human speech in more detail than anyone before him, and how the various organs involved in speaking might be emulated by a machine. He outlined his own attempts to build such a machine, alluding to the many blind alleys he encountered along the way, and noting that "I have thrown away so many components, that a strong horse could scarcely draw them away." Finally, he described in detail the design of his most advanced speaking machine, with the intention of making it easy for other researchers in the field to construct their own versions and devise further improvements.

Kempelen's book established him as the founder of the discipline known today as experimental phonetics, and his investigation of the mechanism of speech was the most advanced of its time. His final speaking machine survives to this day in the Museum of Science and Technology in Munich. A copy of this machine was built in the nineteenth century by Charles Wheatstone, the English scientist remembered for pioneering the electric telegraph. In 1863 Wheatstone gave a demonstration of his speaking machine to a young boy named Alexander Graham Bell. Inspired by what he had seen and heard, Bell immediately set out to build his own speaking machine, and his research into the mechanism, imitation, and transmission of speech ultimately led to his invention of the telephone in 1876.

The speaking machine was evidently impressive, but Kempelen regarded it as unfinished. His original intention, according to Windisch, was to house his speaking machine inside a human figure; Kempelen had in mind the form of a five- or six-year-old child, given the machine's infantile voice. But having failed to make as much progress as he had anticipated in extending his machine's abilities – many sounds and words remained forever beyond its reach – he abandoned this idea. What is notable, however, is that alongside the Turk, which appeared to be capable of reason, Kempelen's speaking machine emulated a uniquely human capability, never before witnessed in a machine.

*

While Kempelen spent countless evenings experimenting in his study, which resounded to the honking and droning of his speaking machines, by day he continued to work as a civil servant and was elevated to the position of privy councillor in 1786. He also pursued many other interests, both scientific and artistic. He built a typewriter that could be used by a blind operator and spent much time, effort, and money building his own kind of steam engine – though, according to one account, it exploded. He also wrote two plays and produced a number of engravings, including several landscapes.

In other words, once his tour of Europe ended in 1785, Kempelen finally succeeded in putting the Turk behind him. Thomas Collinson, an Englishman who visited Kempelen around 1790, was given a demonstration of the speaking machine but noted that "not a word passed about the chess-player; and of course I did not ask to see it." As one of Kempelen's friends observed, his reticence to talk about the chess player was simply because it was the invention "on which he prides himself the least; he often mentions it as a mere bagatelle". But Collinson speculated that the true reason for Kempelen's silence was that the Turk's secret had finally been uncovered in 1789 by Joseph Friedrich, Freiherr zu Racknitz, the author of a book that was unquestionably the most detailed analysis of the Turk since its debut.

Racknitz had seen several of the Turk's performances in Dresden in 1784 and had struck up an acquaintance with Kempelen in order to be able to question him more closely about the automaton. Over the next five years he built a series of models of the Turk to test his theories about how it might work. His book, the title of which translates as "On the chess player of Mr. von Kempelen and an imitation of it", summarized his work and was richly illustrated with engravings of his models – all of which, Racknitz admitted, constituted "a hypothesis, pure guesswork, to explain Mr. Kempelen's chess machine, and an outline of how to build a similar one".

Such was Racknitz's interest in the Turk that he made a thorough review of all previous descriptions of it, including those of Windisch, Decremps, Ebert, Hindenburg, and Böckmann. He also spoke to members of the audience at the Turk's performances to find out why they thought it might work. He then boiled down the various suggested explanations for its secret into five separate hypotheses: namely, that it was a genuine automaton that really could play chess, pre-programmed with the correct response to every possible move in every possible position; that it was a genuine automaton, capable of making a few moves on its own but occasionally guided by a human operator (the Paris theory, favoured by Hindenburg); that it was operated externally by magnets; that it was operated externally by hidden strings; and that it was operated by a person concealed within the machine itself.

So which of these five theories, if any, was correct? "All these hypotheses contain so many flaws," wrote Racknitz, "that to start with I could not agree with any of them." But he then considered each theory in more detail. He dismissed the idea that the Turk was a pure automaton on the grounds that Kempelen would have had to pre-calculate the appropriate move to make in response to every possible move in every possible chess position – something that, he argued, could not possibly have been done within six months, the time Kempelen took to build the Turk. (The idea that the Turk might somehow perform the calculations itself during the game was evidently dismissed as being too outlandish.) Racknitz also pointed out that the machinery inside the Turk did not look complicated enough to respond to so many different moves; and that if the Turk was a pure machine, why did Kempelen not leave the doors open while it played?

Moving on to the second hypothesis, Racknitz raised similar objections: the Turk's machinery did not look complicated enough to enable it to play by itself even for a handful of moves. The idea of magnetism was also dismissed, because Kempelen and Anthon stood too far away from the automaton for magnets hidden in their pockets to have any effect on its internal mechanism. Hidden strings were clearly out of the question because the automaton was freely moved around before play and had made appearances in a large number of venues, not all of which could have been suitably prepared in advance. This left only one hypothesis: that of a concealed operator. But, asked Racknitz, "where was this person during the time that the innards of the machine were exposed to view? How could he follow the development of the game? How is he supplied with necessary light, and air to breathe? And how does he prevent his presence from being revealed through accidental coughs and sneezes?"

Racknitz went on to answer these questions as best he could. He proposed that the drawer beneath the Turk's three doors did not extend all the way to the back of the cabinet; one of the audience members he questioned claimed to have heard the drawer touching something as it was opened. This, he suggested, might leave enough room at the back of the cabinet, beneath the floor of the main compartment, for someone to hide. Once the main compartment had been shown to the audience and the doors closed, the operator could then sit up (his feet hidden behind the machinery in the Turk's small compartment, and his head just beneath the chessboard overhead) and operate the Turk's arm using a set of levers. The operator followed the game, Racknitz suggested, by watching the movement of magnetized needles suspended beneath the squares of the chessboard. Each chessman, he assumed, had a magnet inside it, which would affect the position of these needles as it was moved from one square to another – an idea Racknitz

This engraving and the two following from Racknitz's book show how he thought the Turk worked. First the interior of the cabinet was revealed.

claimed to have had independently of Böckmann. Holes in the bottom of the main compartment provided air; two candles provided light for the operator to follow the game on a small secondary chessboard. And to avoid embarrassing coughs and sneezes, the operator would not play when suffering from a cold or would use the noise of the clockwork mechanism to conceal any involuntary sounds.

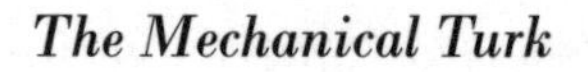

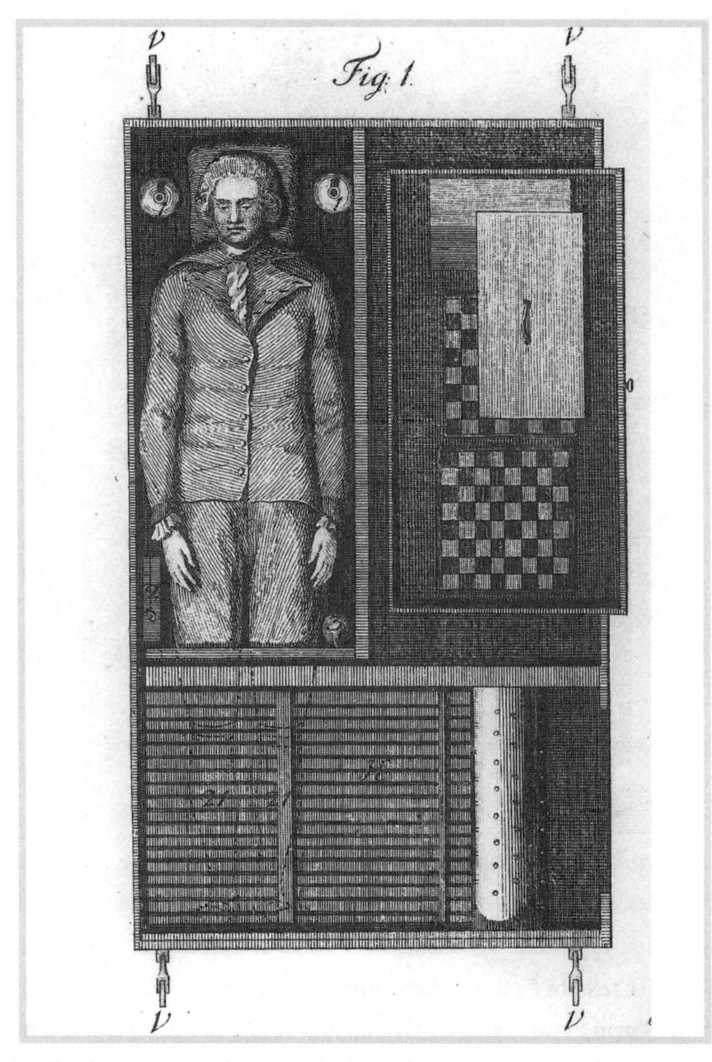

Racknitz assumed that while the interior of the cabinet was being displayed, a diminutive operator lay hidden behind the drawer.

Once the cabinet doors were closed, the operator sat up and worked the automaton.

To test his ideas, Racknitz built a model of the Turk, measuring twelve inches by seven inches by seven inches; he also made separate models of the Turk's mechanical hand and the chessboard, together with a magnetized chess piece (a bishop). He concluded that in every respect, his theory was plausible. It seemed he had found a way to construct a machine that could perform like the Turk; but had he discovered how the Turk itself actually worked? Eager to find out whether he had guessed correctly, Racknitz sent a copy of his book to Kempelen; but according to Collinson, Kempelen was "unwilling to acknowledge that [Racknitz] has completely analysed the whole".

There are, indeed, a few problems with Racknitz's explanation. First, his model was not in proportion to the actual Turk; the cabinet was far too long in relation to its height and depth. Second, even according to Racknitz's distorted measurements, the operator hiding behind the drawer would have to have fitted into a space five feet long, eighteen inches wide, and about seven inches high – surely an impossibility for an adult. Racknitz's engravings show a diminutive operator smaller than the Turkish figure itself, which Racknitz described as medium-sized. Although he does not say so explicitly, Racknitz seems to be implying that the operator was either a dwarf or a small child. But other than saying that this explanation was incorrect, Kempelen refused to be drawn out on the extent to which Racknitz had solved the puzzle of the Turk's operation. Kempelen, it seems, was determined to take the secret with him to his grave.

CHAPTER SIX

Adventures of the Imagination

THE RUSSIAN DEFENSE (E4 E5 NF3 NF6): An opening in which Black counterattacks instead of defending.

Many must be the adventures of the Automaton, lost, unhappily, to the knowledge of man. A being that kept so much good company, during so long a space of time, must, indeed, have gone through an infinity of interesting events. In this age of autobiography, when so many wooden men and women have the assurance to thrust their personal memoirs on the world, a book on the life and adventures of the Automaton Chess-player would surely be received with proportionate interest.

– George Walker, "Anatomy of the Chess Automaton" (1839)

Although the Turk did nothing more than gather dust in Schönbrunn palace in Vienna for nearly twenty years between 1785 and 1804, a number of stories purport to recount its adventures during this period. In 1785, for example, Kempelen is said to have taken his automaton to Berlin at the invitation of King Frederick II of Prussia, better known as Frederick the Great. The king was reputed to have played a game of correspondence chess (in which the players communicate their moves to each other by letter) with Voltaire in Paris. So it seems plausible that Frederick might have invited Kempelen to his court to demonstrate his automaton.

The story goes that the Turk duly took on and defeated the court's finest players, including Frederick himself, who was suddenly consumed with a desire to discover the automaton's secret. Being a king, he used an approach that differed from that of a scientific man or a pamphleteer: instead of getting involved with evaluating the various theories that had been proposed, Frederick simply offered to buy the automaton for an enormous sum, so that he could examine its interior directly. Having handed over the cash, Frederick was let in on the automaton's secret by Kempelen and was horrified to discover that he had been fooled by a very simple trick. Frederick never revealed the secret but ordered the automaton packed up and consigned to an obscure corner of his palace.

This tale, while outwardly believable, is undermined by the fact that the Turk undoubtedly remained in Kempelen's possession. Furthermore, there is no documentary evidence for it. Frederick was certainly interested in automata, but he does not seem to have had a particular interest in chess. According to one biographer, he admitted to his nephew that he knew nothing about the game at all, so although Frederick did correspond with Voltaire, the correspondence game seems unlikely. Most suspiciously of all, the tale of Frederick's encounter with the Turk first appeared in the early nineteenth century, by which time the Turk was also credited with having played against King George III of England. Not one of the accounts of the Turk's exploits during the 1780s mentions Frederick or George, so it seems very unlikely that the Berlin episode or the Turk's supposed match against George III ever took place.

A far more elaborate story was that told by the great nineteenth-century French magician Jean Robert-Houdin in his memoirs, which were published in 1858. Given his own interest in automata and their close relationship with conjuring, it was only natural for Robert-Houdin to include an account of the Turk's origins, which he claimed to have been told by the nephew of a Russian doctor named Osloff. Like many supposedly historical accounts from the mid-nineteenth century, Robert-Houdin's story contains lengthy passages of dialogue; but reconstructing (that is, fabricating) conversations between historical figures for the purposes of livening up a book or newspaper article was a common practice at the time.

According to Robert-Houdin, Kempelen made a trip to Russia to learn the language as part of his research into the theory of speech. While in Russia, he went to stay with his friend Osloff, only to discover that the doctor was sheltering a fugitive army officer named Worousky. Worousky, who was Polish, had led a rebellion by Polish soldiers serving in the Russian army. In the ensuing battle, his legs were shattered by a cannonball, and Worousky escaped the massacre of his fellow rebels only by hiding in a ditch. He then dragged himself to Osloff's house. The doctor took pity on him but was unable to save Worousky's legs, which had to be amputated. During his convalescence, Worousky played many games of chess with Osloff and revealed himself to be an unusually cunning and gifted player.

By the time of Kempelen's arrival at Osloff's house, a price had been put on Worousky's head by the Russian authorities, and the doctor was anxiously searching for a way to spirit his patient out of the country. It was at this point, says Robert-Houdin, that Kempelen had the idea of building a chess-playing automaton in the form of a wooden figure seated behind a cabinet. Keeping this scheme to himself, he worked day and night and completed its construction in a mere three months. The automaton then played its first game, against Osloff. Once the game was over, Osloff was astonished to discover that he had in fact been playing against Worousky, who was hiding inside the automaton. Since he had no legs, Worousky could fit easily inside the cabinet, evading detection by changing his position as the various doors were opened and shut, so that he ended up inside the wooden figure, with a good view of the chessboard. Kempelen agreed to smuggle Worousky out of Russia by mounting a tour with the automaton, stopping at a handful of towns on the road to the border.

The party set off, and at its first stop, in the town of Tula, the automaton beat everyone who challenged it and caused general amazement. It then continued on its way, stopping at Kaluga, Smolensk, and Vitebsk, its fame growing all the while. But just as he was nearing the border and it looked as though his plan would succeed, Kempelen received a letter ordering him to go directly to the imperial palace in St. Petersburg. Catherine the Great, empress of Russia, had heard about the automaton and wished to see it for herself. Kempelen was distraught, but Worousky, who was now one of the most wanted men in Russia, relished the idea of playing chess while hiding right under the empress's nose. The two men duly proceeded to St. Petersburg and set up the automaton in the royal library. That evening, Catherine herself challenged it to a game.

The automaton played particularly aggressively, and the empress quickly lost a bishop and a knight. She then tried to make an illegal move, which the automaton immediately corrected. Catherine made the same illegal move again, at which point the automaton swept all the pieces off the board, ending the game. Delighted, the empress offered to buy the automaton, so that she would always have such a spirited opponent close at hand. Kempelen, however, insisted that he could not sell the automaton. He explained that its operation relied on a trick, and that it could not therefore perform without him. The empress accepted this explanation and, never suspecting the presence of a man inside the machine, sent Kempelen and his automaton on their way. Worousky thus escaped to England, where Kempelen continued to display his automaton.

This story is undoubtedly enchanting, but it is also entirely untrue. Aside from its other implausible aspects, the chief problem is the date, for according to Robert-Houdin all of this is supposed to have happened in 1796. Yet it is certain that Kempelen actually built his automaton nearly thirty years earlier, and that it made its debut in Vienna; there is also incontrovertible proof of its travels in Europe during the 1780s. Furthermore, Robert-Houdin has the automaton playing his first game on October 10, 1796, just a month before Catherine's death, by which time she was gravely ill. Another problem with Robert-Houdin's tale is that the automaton's supposed itinerary would have taken it deeper into Russia, rather than toward western Europe. So despite the presence of one or two accurate details – there was indeed an uprising by Polish soldiers in the 1790s, for example, and Kempelen's automaton did always correct illegal moves – Robert-Houdin's entire account must be dismissed as fiction.

Even so, the tale of the Turk's encounter with Catherine the Great remains one of the most widely told stories about the automaton, and the notion that Kempelen's invention was operated by a legless Pole named Worousky has often subsequently been presented as a fact. The 1911 edition of *Encyclopædia Britannica,* for example, claims that the Turk was operated by "a Polish patriot, Worousky, who had lost both legs in a campaign; as he was furnished with artificial limbs when in public, his appearance, together with the fact that no dwarf or child travelled in Kempelen's company, dispelled the suspicion that any person could be employed inside the machine."

Robert-Houdin's version of the story inspired several subsequent plays and books, all of which added further embellishments. *La Czarine,* a play first staged in Paris in 1868, added a romantic subplot in which the Polish patriot, now renamed Vorowski and with his legs restored, fell in love with Kempelen's daughter. Robert-Houdin himself provided the special effects for this production, in which Kempelen's technological wizardry was depicted through the use of illusions and conjuring tricks. Sheila E. Braine, author of *The Turkish Automaton,* a novel published in 1894, added a few further twists of her own by introducing Kempelen's son and Orloff's children as additional characters. In her version, Kempelen and Worousky visit Paris after their escape, and the automaton defeats Philidor. The 1926 novel by Henry Dupuy-Mazuel, *The Chess Player,* is less cheerful. It culminates in a scene in which Catherine orders the automaton to be executed by firing squad, killing the person hidden inside it. But all of these tales were fictionalized versions of a story that was itself entirely without foundation, for the meeting between Kempelen's automaton and Catherine the Great never took place.

*

One event that definitely did occur around this time was a fateful visit made by a young boy to the London museum of John Joseph Merlin, the celebrated inventor and automaton maker. As well as opening his Mechanical Museum to the public during the day, Merlin kept it open in the evenings for the benefit of "young amateurs of mechanism" whom he hoped to inspire. One such youngster was a Devonshire boy named Charles Babbage, whose mother took him to Merlin's museum around 1800, when he was about eight years old. Babbage wrote in his memoirs that from a very young age, "my invariable question on receiving a new toy, was Mamma, what is inside of it?"

The young Babbage was particularly interested in the automata on display, so Merlin took him behind the scenes to his workshop to show him some further examples. In the workshop, Babbage later wrote, "there were two uncovered female figures of silver, about twelve inches high." The first was a brass clockwork figure, able to perform "almost every motion and inclination of the human body, viz. the head, the breasts, the neck, the arms, and fingers, the legs & c., even to the motion of the eyelids, and the lifting up of the hands and fingers to the face . . . she used an eye-glass occasionally and bowed frequently as if recognising her acquaintances."

It was the second automaton figure, however, that really caught Babbage's eye: a dancing lady, "with a bird on the forefinger of her right hand, which wagged its tail, flapped its wings and opened its beak". Babbage was spellbound: "The lady attitudinized in a most fascinating manner. Her eyes were full of imagination, and irresistible." It was probably this automaton's graceful and lifelike motion that first introduced Babbage to the fanciful possibility that mechanical devices might, in some respects, be able to emulate human abilities. A few years later Babbage was to encounter the Turk itself, before going on to become famous for designing a series of mechanical computers – elaborate machines capable of emulating the uniquely human capacity for logical calculation. These machines, like the Turk, would be the subject of arguments about the possibility of machine intelligence. Kempelen's automaton was also to inspire Babbage to investigate the possibility of building his own chess-playing machine. In other words, as a result of Babbage's visit to Merlin's Mechanical Museum, the subjects of chess, intelligence, and computing were well on their way to becoming inextricably intertwined.

*

Even while it lay dormant in Vienna during the last years of Kempelen's life, the Turk continued to be the source of speculation and storytelling across Europe. In addition to inspiring tall tales of its encounters with European heads of state, the Turk was the inspiration for fiction of the more conventional kind. *The Chess Machine,* a play written by the German actor Heinrich Beck, was published in Leipzig in 1797 and went on to become quite popular in the early nineteenth century, with further editions published in Berlin, Vienna, Milan, and Venice.

Another drama, *The Chess Player,* by a Frenchman named Benoit-Joseph Marsollier, played in Paris in 1800 and 1801. Its plot involved a young soldier who fell in love with the daughter of an elderly chess player and surreptitiously entered his house hidden inside a chess automaton, whereupon the young woman suddenly developed a keen interest in chess. The story ended happily, with the couple's marriage.

Evidently Kempelen's automaton, or at least the dramatic possibilities of a person concealed within a chess-playing machine, had touched a nerve. The Turk had become a living legend, despite Kempelen's efforts to dampen public enthusiasm for it.

After Joseph II's death in 1790, Kempelen continued to serve under his successor, Leopold II, and then under Leopold's successor, Francis II, who ascended to the imperial throne in 1792. Kempelen retired in 1798 and spent the remaining years of his life living in Vienna, where his house became a popular meeting place for scientists, artists, and other intellectuals. In 1801 Kempelen was offered a senior post at the Royal Academy of Art in Vienna by Count Cobenzl (who had been defeated by the Turk at its debut many years earlier), but he declined on the grounds of ill health.

Wolfgang von Kempelen died on March 26, 1804, at the age of seventy, and was buried two days later in Vienna. Kempelen's family later erected a memorial in his name, on which were engraved the words of the Roman poet Horace: *Non omnis moriar,* which means "I do not die completely". This epitaph was entirely appropriate, for Kempelen's ingenious spirit lived on, though not perhaps in the way he would have chosen. Kempelen had tried hard in the last years of his life to ensure that his chess-playing automaton would not overshadow his other achievements. But within a few years of his death the indefatigable Turk – packed into his wooden crates, with the principle of its operation still an unsolved mystery – was to experience a spectacular resurrection.

CHAPTER SEVEN

The Emperor and the Prince

NAPOLEON'S RETREAT FROM MOSCOW: A chess puzzle in which Napoleon, represented by the black king, must cross the board, which is dominated by the white queen controlling an entire diagonal. The black king is pursued by the two white knights, representing the cossack cavalry. This puzzle first appeared in a book published in St. Petersburg in 1824.

Chess is war over the board. The object is to crush the opponent's ego. – Bobby Fischer

In May 1809 Napoleon Bonaparte, the emperor of France, marched into Vienna and established his headquarters at Schönbrunn palace. At the time Napoleon was the most powerful man in Europe. His empire stretched from France, northern Italy, and the low countries across most of central Europe, including Austria and Hungary. But in February 1809 the Austrian army had taken advantage of Napoleon's preoccupation with the invasion of Spain to stage a rebellion. Napoleon returned to Austria, and in July he defeated the Austrian army in a two-day battle of shocking severity at Wagram. He then spent the summer at Schönbrunn while a peace treaty was negotiated.

All kinds of entertainment were put on during this period, including concerts, ballets, theatrical performances, firework displays, and, in August, an elaborate birthday celebration for the emperor. Among the many performers, artists, and scientists to appear before Napoleon over the course of the summer was Johann Nepomuk Maelzel, an inventor. He displayed a number of artificial limbs of his own design, intended for soldiers who had lost limbs in battle. Napoleon was impressed by what he saw, and suggested that Maelzel design a collapsible cart for transporting the wounded from the battlefield, akin to the folding gurneys used by modern ambulance crews. Maelzel agreed and then mentioned that he happened to have another piece of machinery that he thought might interest the emperor: a chess-playing automaton.

*

Maelzel had bought the Turk from Kempelen's son a few years earlier, and there could surely have been nobody in Europe better qualified to act as its guardian. Born in Bavaria in 1772, Maelzel was the son of an organ builder and emerged from his upbringing skilled as both an engineer and a musician. After a spell working as a piano teacher, he traveled to London and Paris to study mechanical engineering and then set about combining his two skills through the construction of a musical automaton on a grandiose scale. As an engineer, Maelzel was acclaimed as the equal of Vaucanson and Kempelen; but his greatest talent was as a showman with an uncanny feel for public taste, which prompted one writer to describe him as the "prince of entertainers". He called his musical automaton, first displayed in Vienna in 1805, the Panharmonicon. It consisted of an entire mechanical orchestra, including trumpets, clarinets, violins, cellos, and percussion, all packed into a machine measuring six feet wide, six feet deep, and five feet tall, and controlled by a rotating studded drum, like an enormous music box.

The Panharmonicon was well received by the public and was particularly admired for the subtlety of its playing of quiet passages. In early 1807 Maelzel took the Panharmonicon to Paris, where it gave two performances a day for several months. In July Maelzel extended the performances with additional works and doubled the price of entrance from three to six francs. Leaving the Panharmonicon in Paris in the care of a manager, he then returned to Vienna and reappeared in Paris the following year with a new creation: an automaton trumpet player.

The trumpet player was even more remarkable than the Panharmonicon. Indeed, it was widely deemed to be even more impressive than Vaucanson's two musical automata, and its playing was so lifelike that many observers wrongly assumed it was a fraud. By changing the notched drum inside the trumpeter's chest, Maelzel could program it to play different pieces. The trumpet player usually performed cavalry marches and was often accompanied by Maelzel himself at the piano.

In 1808 Maelzel sold the Panharmonicon for a reputed 60,000 francs and then returned to Vienna, where he was appointed engineer to the royal court at Schönbrunn, a post similar to the one once held by Kempelen. The two men probably knew each other, if only vaguely; at one stage Maelzel may even have tried to buy the Turk from Kempelen. Eventually, after Kempelen's death, Maelzel bought the Turk from his son. He then had to work out how to assemble it, to repair any broken components, and to rediscover the secret of its operation for himself. But he had done all of this by 1809, and the Turk was in full working order and ready for display when Napoleon arrived at Schönbrunn that year.

*

Napoleon's encounter with the Turk is easily the most famous episode in its career. According to the eyewitness account of Napoleon's valet, Louis-Constant Wairy, Maelzel claimed to have built the chess player himself. This would not be the last time that Maelzel tried to pass off someone else's invention as his own; but the Turk had by this point been out of circulation for a quarter of a century, so this claim was unlikely to be contradicted. Besides, Maelzel subsequently made several improvements to the automaton's mechanism in order to make its secret even more difficult to fathom, so he may have felt justified in referring to it as his own creation.

Maelzel set up the Turk in the apartment of the Prince de Neufchâtel, one of Napoleon's most trusted generals. "The Emperor went there, and I followed him with several other persons," recalled Constant (as he was known) in his memoirs, which were published in 1830. There are numerous accounts of Napoleon's match with the Turk, some of

Napoleon Bonaparte

which are more credible than others. Although the memoirs of Constant were ghostwritten and contain a number of highly implausible tales, his account of the match is brief and straightforward. Other, more embellished versions of the story seem to have built on Constant's foundations but then go off in different directions, which would explain why they are contradictory.

"The automaton was seated in front of a table on which a chessboard was arranged for a game," wrote Constant. "His Majesty took a chair, and sitting down opposite the automaton, said, laughing: 'Come on, comrade; here's to us two.' The automaton saluted and made a sign with the hand to the Emperor, as if to bid him begin.

"The game opened, the Emperor made two or three

moves, and intentionally a false one. The automaton bowed, took up the piece and put it back in its place. His Majesty cheated a second time; the automaton saluted again, but confiscated the piece. 'That's right,' said His Majesty, and cheated the third time. Then the automaton shook its head, and passing its hand over the chessboard, it upset the whole game. The Emperor complimented the mechanician highly."

The theme of Napoleon daring the automaton to challenge his authority when he breaks the rules is central to all the accounts of his encounter with the Turk. According to another version of the tale, related by the American chess writer George Allen in 1859, Maelzel did not allow Napoleon to sit at the Turk's own chessboard but adopted Kempelen's practice and insisted that he sit at a separate table with its own chessboard, roped off from the Turk by a silken cord. When Napoleon attempted to walk past the cord, Maelzel is said to have told him, "Sir, it is forbidden to go any further" – at which the emperor merely smiled and returned to his place at the second table. But according to yet another account, by Silas Weir Mitchell, published the same year, Napoleon exclaimed: "I will not contend at a distance! We fight face to face." The Turk nodded, and the emperor sat down in front of the automaton.

Neither of these embellishments to Constant's account can be trusted. Many of the additions to the story of Napoleon's encounter with the Turk seem to have been made in order to provide a fuller description of the automaton to those unfamiliar with it. In the case of Napoleon and the silken cord, such details were probably added to show that nobody – not even the ruler of most of Europe – was usually allowed to approach the Turk during a game. Similarly, Mitchell's version of the story goes on to describe Napoleon, after losing one game to the Turk, placing a large magnet on the cabinet in an attempt to subvert its internal machinery, but losing the next game anyway. Napoleon is then said to have wrapped a lady's shawl around the head and body of the Turk, to test the theory that a hidden operator might be watching the board through an opening in the Turk's chest. Once again, the Turk won the subsequent game, prompting Napoleon to sweep the chessmen off the board and storm out of the room. Again, these details are not credible and seem intended merely to explain to the uninitiated that the Turk still worked with a magnet sitting on the chessboard or a shawl wrapped around its body.

Mitchell's version of the tale, which is undoubtedly the most elaborate, is further undermined by a number of other incorrect claims. He credulously recounts the bogus tale of Frederick the Great's purchase of the Turk and its subsequent storage in Berlin and then asserts that the match with Napoleon took place in that city, which fell to Napoleon in October 1806. Mitchell also spuriously claims that the Turk played against George III of England and even Louis XV of France. (Louis XV died in 1774, by which time the Turk had not ventured much beyond Vienna.) In short, Mitchell's version of events is entirely untrustworthy, as Allen pointed out within a few months of its publication. Allen's more straightforward account, which he claimed to have heard independently from two of Maelzel's acquaintances, is rather more credible; apart from the remark about the silken cord, it agrees closely with the version of the story given by Constant.

The notion that Napoleon encountered the Turk in Berlin seems to have arisen from an anonymous article – published in France in 1834 in the *Magazine Pittoresque* – that was evidently based on information provided by one of Maelzel's associates. This article formed the basis for a longer piece by Mathieu-Jean-Baptiste de Tournay that appeared in *Le Palamède,* the world's first chess magazine, in 1836, and for an article by George Walker, "Anatomy of the Chess Automaton", that was published in *Fraser's Magazine* in 1839. Aside from the erroneous reference to Frederick the Great and Berlin, these three articles make no mention of silken cords, shawls, or magnets, and all three are consistent with Constant's version of the tale.

A further twist to the Napoleon tale appeared in 1844 in the chess column of the *Illustrated London News.* In this version of the story, Napoleon cheated and forced the Turk to sabotage the game, just as Constant describes; but he then asked to play a second game and promised to abide by the rules. This game, which lasted nineteen moves, was supposedly recorded by an eyewitness and is reproduced in the magazine. It is a curious game, for a number of reasons. For starters, the Turk evidently played rather badly, missing an obvious winning opportunity after fifteen moves, choosing instead to capture Napoleon's queen, but eventually securing victory. Worse, no origin for the game is given, so that it must be dismissed as spurious.

Why did it take over twenty years for the first accounts of Napoleon's encounter with the Turk to appear in print? The lack of contemporary reports is not that surprising, since the game would have been witnessed by very few people, and amid the grand drama of Napoleon's career would not have been regarded as a terribly important or noteworthy event – it merits a single paragraph in Constant's four volumes of memoirs. Furthermore, Constant had good reason to wait until 1830 before publishing his memoirs. Between 1815 and 1830 France was ruled by Napoleon's enemies, who effectively prevented the publication of works sympathetic to the emperor by his followers.

In the story's favour it must be said that Napoleon, unlike most of the other heads of state who supposedly played chess with the Turk, was genuinely interested in the game. Napoleon was born, coincidentally, in 1769 – the same year that Kempelen began building his automaton. As a young man in the 1790s, Napoleon played chess in the Café de la Régence in Paris. (The proprietor subsequently had a small marble table, at which Napoleon had played in 1798, inscribed with a commemorative engraving.) There are several anecdotes about Napoleon's chess-playing abilities, though as with the details of his match against the Turk, it is difficult to distinguish fact from fiction. But it is fairly clear that Napoleon was a bad loser and considered himself to be a far better player than he actually was, because his associates feared the consequences if they beat him, and tended to let him win.

The other factor that lends credibility to the whole tale is the indisputable fact that the Turk was purchased by Napoleon's stepson, Eugène de Beauharnais, sometime between 1809 and 1812. Eugène was the son of Napoleon's wife, Josephine, by her previous husband (who had been guillotined in 1794). Although Eugène initially resented his mother's decision to remarry, he soon became very close to Napoleon, who adopted him as his aide-de-camp. Eugène went on to distinguish himself as one of Napoleon's bravest and most devoted followers. By 1809 he had been appointed viceroy of Italy, and Napoleon had arranged his marriage to Princess Amalie Auguste of Bavaria.

Like his stepfather, Eugène was an enthusiastic chess player. Yet although Eugène spent much of the summer of 1809 at Schönbrunn, there is no mention of his having attended Napoleon's match against the Turk. But somehow he heard about Maelzel's automaton and arranged to see it for himself. By the beginning of October it was clear that the peace treaty would soon be signed, and Eugène, realizing he would soon be on his way back to Milan, went on a shopping spree. He bought toys for his children and porcelain and fine engravings for his wife. He also bought some larger items, including a number of horses and at least two pianos. Another large item on his shopping list that month may well have been the Turk.

Eugène was so intrigued by the automaton that he insisted on being told its secret. Maelzel would only divulge it if Eugène agreed to buy the automaton, and when it became clear that Eugène was serious, Maelzel named his price: 30,000 francs, or three times what Maelzel himself paid for the machine. Eugène agreed and was then let in on the Turk's secret. And this, it seems, is the origin of the stories that other heads of state – including Frederick the Great, Catherine the Great, and, inevitably, Napoleon himself – tried to buy the Turk to discover its secret. Such tales appear in books and articles about the Turk only after Eugène's acquisition of the automaton.

Although Eugène definitely bought the automaton, it is not certain that the sale took place at Schönbrunn in 1809; another possibility is that Eugène arranged for Maelzel to visit him in Milan and bought the Turk from him on that occasion. In any case, the automaton ended up in Eugène's possession. Eugène subsequently lost interest in his purchase; indeed, he may have felt duped by Maelzel. In 1812 the Turk was seen by Carl Bernhard, the duke of Saxe-Weimar, in a state of "inglorious repose" in Eugène's palace in Milan, and by another traveller named Millin at around the same time. That same year Eugène followed his step-father on his disastrous Russian campaign. Neglected and forgotten, the Turk went into another of its many periods of hibernation.

*

Maelzel spent most of 1812 working on a new, improved Panharmonicon; it is even possible that he sold the Turk to fund its construction. Maelzel bought and sold automata with abandon throughout his life, as well as building his own. But his finances were usually rather haphazard, since his talent for earning money quickly through his mechanical displays was offset by an equal and opposite talent for profligate spending, on fine food and wine in particular. Maelzel had high hopes for his new Panharmonicon – namely, that it would outshine his previous creations and become the talk of Europe.

Having established his workshop in a piano factory in Vienna, he struck up a new friendship with the composer Ludwig van Beethoven. While working on the Panharmonicon, Maelzel was also developing a number of other inventions, including a set of ear trumpets for Beethoven (who had been losing his hearing for many years) and a device called a musical chronometer. The latter consisted of a toothed wheel that caused a lever to rise and fall, striking a small piece of wood and making a regular ticking sound.

The rate of the ticking was adjustable, so that the device could be used to measure the tempo of a piece of music; it was thus a forerunner to the metronome.

But Maelzel's chief concern was his new Panharmonicon, which was ready to go on public display by the end of 1812. At seven feet tall, it was even larger than the original, with an expanded percussion section, more string instruments, and a wide repertoire of music by Haydn, Handel, and Cherubini, encoded onto studded cylinders. It was given pride of place in Maelzel's new *Künstkabinett* (literally, "art room"), a museum of artistic and scientific curiosities by Maelzel and others. On display were marbles, bronzes, paintings, and automata. But the stars of the show were undoubtedly Maelzel's automaton trumpeter, playing a cavalry march and accompanied by Maelzel at the piano, and the new Panharmonicon. Maelzel began planning a trip to London, where he thought his two musical automata would be most appreciated and hence most profitable.

By this time Maelzel had started work on another elaborate piece of machinery: a mechanically animated panoramic painting, or diorama, titled the Conflagration of Moscow, depicting the destruction of the city in September 1812 by its residents, who chose to burn it down rather than let it fall into the hands of Napoleon's advancing army. Maelzel knew his diorama would delight the citizens of Vienna, whose city had twice fallen to Napoleon, and who now relished the reversal in the emperor's fortunes. Similarly, Maelzel knew his diorama would go down very well in London, whose inhabitants also despised Napoleon. So he decided to delay his trip until the diorama, with its tiny models of houses, churches, and bridges, and an elaborate system of sound effects, musical accompaniment, and pyrotechnics, was complete.

On June 21, 1813, the French army was defeated by the duke of Wellington at Vitoria, decisively ending Napoleon's dream of annexing Spain. News of the victory inspired Maelzel to hatch a plan: he would commission a new piece of music for the Panharmonicon to mark the occasion, which would be guaranteed to cause a sensation at its first performance in London. At the time, battles and sieges were popular subjects for musical works, typically featuring drum marches, trumpet flourishes, and rousing national themes. Maelzel jotted down the rough outline of such a piece and asked Beethoven to write it for him. The composer readily agreed. In 1804 Beethoven had dedicated his third symphony, the Eroica, to Napoleon; but when Napoleon proclaimed himself emperor the following year, Beethoven erased the dedication in disgust.

Within four months Beethoven had completed the score. Maelzel started transferring the new piece, titled "Wellington's Victory", onto studded Panharmonicon cylinders, and as he did so he had another idea. Beethoven wanted to accompany him on the voyage to London but was short of money (in fact Maelzel loaned him fifty ducats); so Maelzel proposed putting on an orchestral performance of the new work in Vienna to raise money for the trip. The programme would consist of another new composition by Beethoven (his seventh symphony), Maelzel's automaton trumpeter playing two marches accompanied by a full orchestra, and finally the orchestral arrangement of "Wellington's Victory". The Panharmonicon would have no part in this concert, but Maelzel hoped that a performance of the new piece by a full orchestra would increase the subsequent demand for the Panharmonicon version.

Beethoven and Maelzel called in favours and exploited their connections within Vienna's musical community and quickly assembled an orchestra consisting of many of the city's finest musicians. The concert, which took place on December 8, was so successful that a second performance was staged on December 12. Beethoven then organized a third performance for January 2, 1814. But by this time he and Maelzel had fallen out. At issue was the question of the ownership of "Wellington's Victory". Maelzel considered it his property; as he saw it, he had outlined the piece, and Beethoven had filled in the blanks as a "friendly gift" to him. But Beethoven took offence at the posters Maelzel had printed to publicize the concert, since they did not acknowledge him as the composer of the new work. He claimed that Maelzel had tricked him into selling the rights to the piece for fifty ducats. The two men went to a lawyer and tried to come to an agreement. But given their contrasting temperaments – Maelzel was dashing and flamboyant, Beethoven brusque and abrupt – it is not surprising that they were unable to settle their differences. Maelzel then infuriated Beethoven by giving two performances of the new piece, played by the Panharmonicon, in Munich. Beethoven responded by launching a lawsuit against Maelzel and writing to his friends in London to inform them that Maelzel had no right to perform the new work.

Having closed his *Künstkabinett* museum in preparation for his trip to London, and with his reputation under attack from Beethoven, Maelzel left Vienna and went to Amsterdam, where he resumed work on his musical chronometer. On a previous trip to Amsterdam, Maelzel had visited the Dutch inventor Diedrich Nikolaus Winkel, who was working on a similar device. Maelzel went back to visit Winkel in 1815 to see how he was getting on, and Winkel proudly showed him his new design: a pendulum in the form of a metal rod, whose rate of oscillation could be controlled by adjusting the position of a sliding weight. Maelzel realized that this approach was far superior to his own design and offered to buy the rights to it, but Winkel refused to sell them. So in 1816, after adding a couple of refinements of his own, Maelzel went to Paris and, passing off Winkel's invention as his own, applied for a patent. He then established a company to manufacture and distribute this new device, which he dubbed the "Maelzel Metronome". Over the next few months the metronome sold well in France, Britain, and

America but did not take off in Germany or Austria. In 1817 Maelzel decided to return to Vienna to settle his differences with Beethoven and ask him to endorse the metronome.

Surprisingly, Maelzel and Beethoven quickly forgot about their feud and agreed to divide the legal costs between them. Beethoven was initially sceptical about the metronome – "one must feel the tempos," he declared – but changed his mind and started marking his scores with "M.M." (for Maelzel Metronome) and a number to indicate the correct tempo setting. Yet despite this valuable endorsement, Maelzel was, somewhat characteristically, soon looking around for a new project to occupy himself. Evidently he had had enough of musical machinery, for he decided that the time had come to renew his acquaintance with an old friend: the Turk.

The automaton was by this time in Munich, where it had been taken by Eugène de Beauharnais, who had retired there after Napoleon's final defeat at Waterloo in 1815. So Maelzel travelled to Munich to negotiate its repurchase. The exact terms of the deal he struck with Eugène, who was now styled the duke of Leuchtenberg, are not entirely clear. Evidently Maelzel was short of funds, for despite his many money-spinning schemes, he never held on to his riches for very long. But Eugène insisted that he would only sell the Turk for the same price he had paid for it, namely, 30,000 francs.

According to one account (de Tournay's 1836 article in *Le Palamède*), Maelzel agreed to pay the interest on this sum, as a form of rent, from the profits of displaying the automaton, which would remain Eugène's property. But an associate of Maelzel's claimed that Maelzel agreed to buy the Turk and to pay the full amount in instalments. In a letter written several years later Maelzel referred to the Turk as "the automaton chess-player entrusted to me by Prince Eugène," which implies that it did indeed remain Eugène's property. Whatever the financial arrangement, however, Eugène handed over the Turk, on one further condition: Maelzel was to display the automaton only within Europe.

Having restored the Turk to full working order, Maelzel headed to Paris, which was still the chess capital of the world. For most of 1818 he exhibited the Turk alongside the automaton trumpeter and the Panharmonicon. But he still dreamed of taking the Panharmonicon and the Conflagration of Moscow to England. By the autumn he had accumulated enough money to fund the journey. Maelzel packed up the automaton once more and took his multifarious machines to London – where the Turk was to encounter by far the most perceptive investigator of its still-mysterious mechanism.

CHAPTER EIGHT

The Province of Intellect

ESCAPE SQUARE: A square vacated by another chessman to enable the king to escape being checkmated.

Thou wondrous cause of speculation,
Of deep research and cogitation.
Of many a head and many a nation,
While all in vain
Have tried their wits to answer whether,
In silver, gold, steel, silk or leather,
Or human parts, of all together,
Consists thy brain!

– Hannah Flagg Gould,
from "Address to the Automaton
Chess Player" (1826)

Just as Johann Maelzel had anticipated, his collection of automata found a receptive and enthusiastic audience in London. To start with, Maelzel rented a small space at No. 4, Spring Gardens – which had remained the traditional venue for exhibitions of automata in London – where he put the Turk and the automaton trumpeter on display. It was the trumpeter that first attracted attention, due to its refreshing humility and consistency when compared with human performers. The *Literary Gazette* of September 26, 1818, noted with approval that when asked to play an encore, the trumpeter "displayed none of the airs of inflated genius, but readily submitted to be prompted, alias wound up, repeated the tunes with the same brilliant execution, and without introducing new ornaments to spoil what had given so much satisfaction before. We commend this example of modesty, equanimity, and propriety to all our first-rate singers and musicans."

But soon the Turk was back in the public eye. It was winning almost every game it played, attracting a great deal of favourable press coverage, and drawing large crowds. Maelzel initially presented his exhibition three days a week; the chess player appeared at 1 P.M. and 3 P.M., with a longer evening performance at 8 P.M., heralded with a performance by the trumpeter. By February 1819 demand was such that Maelzel was opening his doors every day of the week except Sunday. In the cramped surroundings of No. 4, Spring Gardens, he let the Turk's opponents sit right in front of the automaton and did not use a second chessboard on a separate table.

Indeed, Maelzel displayed the Turk in exactly the same manner as Kempelen had previously, with one exception: he dispensed with the mysterious wooden casket into which Kempelen had occasionally peered during performances and which, he sometimes implied, had some sort of magical role in animating the Turk. According to the chess writer Richard Twiss, who had asked Kempelen about this during the Turk's original visit to London in 1783, "a small square box during the game, was frequently consulted by the exhibitor; and herein consisted the secret, which he told me he could in a moment communicate." In truth, however, Maelzel's decision to do away with the casket made no difference one way or another to the mystery of the Turk's mechanism, since none of the proposed explanations referred to it as anything other than a decoy. When quizzed on the subject, Maelzel said simply, "The people are now intelligent; then they were superstitious."

One mark of the degree of interest in the Turk was the

appearance of a new pamphlet describing its performances and speculating about how it might work. This pamphlet, *Observations on the Automaton Chess-Player, Now Being Exhibited in London,* authored anonymously by "an Oxford graduate", appeared in the spring of 1819 and was widely quoted and reprinted in newspapers and magazines. But although this was the first new work devoted to analysis of the Turk since Racknitz's book in 1789, the Oxford graduate was hardly a credit to his university. Much of the pamphlet was simply a rewrite of Windisch's letters, and the feeble speculations that followed – which suggested that the automaton was probably controlled from the outside via "a wire or a piece of catgut, not much thicker than a hair" – championed one of the least credible explanations of the Turk's mechanism, since the Turk was moved around during its performances.

Once the London season ended, the Turk spent the summer of 1819 touring northern England and Scotland, visiting Liverpool, Manchester, Edinburgh, and a number of smaller towns. Returning to London, Maelzel found that the appetite for the Turk had not diminished, thanks in part to the appearance of a new English edition of Windisch's book of letters about it, which had been serialized in at least one newspaper. To accommodate the number of people who now wanted to see the Turk, Maelzel moved his exhibition to a larger venue at No. 29, St. James's Street. This gave him enough room to augment his display with his diorama, the Conflagration of Moscow, and the Panharmonicon, which Maelzel had by this time restyled the Orchestrion. (The change of name was necessary because another mechanical orchestra, shown in London in 1811 and 1817, was also known as the Panharmonicon, a name that had been suggested to its creator by the composer Franz Joseph Haydn.)

Maelzel introduced some new twists to the Turk's presentation during the winter season of 1819–20. First, he enhanced the automaton by adding a simple speaking machine, modelled after those built by Kempelen, enabling it to say "Check" when placing the opponent's king in check rather than nodding three times. Maelzel subsequently modified the speaking apparatus for a trip to France so that the Turk said *"Échec"*, the French equivalent, instead, and thereafter the Turk spoke French.

Furthermore, Maelzel announced that henceforth the Turk would allow its opponent (playing with the black pieces) to move first, thus forfeiting the advantage of having the first move. But that was not all. Maelzel also stated that the Turk would play without one of its eight pawns. Although the pawns are the weakest of the chessmen, this was quite a concession. Yet even with these two handicaps, collectively known as "giving odds of pawn and move", the Turk continued to win almost all its games. Maelzel's new handbill triumphantly announced that "the Automaton Chess Player, being returned from Edinburgh and Liverpool where (giving pawn and move) it baffled all competition, in upwards of 200 games, although opposed by all the best players, has opened its second campaign. . . . Games will be played against any opponent, to whom the double advantage of a pawn and the move will be given."

The handbill declared that the Turk would be displayed together with the automaton trumpeter and the Conflagration of Moscow, which was described as a combination of "the arts of design, mechanism, and music, so as [to] produce, by a novel imitation of nature, a perfect facsimile of the real scene. The view is from an elevated station on the fortress of the Kremlin, at the moment when the inhabitants are evacuating the capital of the czars, and the head of the French columns commences it[s] entry. The gradual progress of the fire, the hurrying bustle of the fugitives, the eagerness of the invaders, and the din of warlike sounds, will tend to impress the spectator with a true idea of a scene which baffles all powers of description." Another notable aspect of the handbill was the announcement in small italic type that "Mr. M[aelzel] begs leave to announce that the Orchestrion, the automaton trumpeter, the Conflagration of Moscow, and the patent for the metronomes, are to be disposed of." Evidently Maelzel needed to raise some money in a hurry, perhaps because he had fallen behind with his payments to Eugène de Beauharnais.

During the 1819–20 season Maelzel arranged for an associate, W. J. Hunneman, to take notes on 50 of the games played by the Turk. These games were then published in a pamphlet, *A Selection of Fifty Games from Those Played by the Automaton Chess-player, During Its Exhibition in London in 1820, Taken Down, by Permission of Mr. Maelzel, at the Time They Were Played.* This pamphlet was then sold at the Turk's exhibitions, providing extra revenue and further publicity. It also enabled would-be opponents of the Turk to study its strategy and playing style before challenging it to a game. Of the 50 games listed, the Turk won 45 and lost 3, with the other 2 games ending in a draw. This was, according to the pamphlet's preface, a representative sample of its playing; in over 300 games played that season, giving pawn and move each time, the Turk was said to have lost only half a dozen or so. In fact, study of the 50 games shows that giving pawn and move may not have been such a concession after all. Removing one of the white pawns meant that even players familiar with the various "standard openings" in chess may have found themselves somewhat disoriented.

The following summer, Maelzel once again took his exhibition to other parts of Britain. He found that brief, limited appearances in particular towns could be extremely lucrative; he claimed that a two-week exhibition in Liverpool was more profitable than six months in London.

In the autumn, as had by now become a regular event, the Turk was back in London for the winter season. It was still receiving plenty of favourable press coverage, no doubt carefully orchestrated by Maelzel. An article in the *New Monthly Magazine* reprinted much of the Oxford graduate's account of the automaton and proclaimed that "this cunning infidel (for he assumes the figure of a Turk) drives kings, and castles, and knights before him with more than moral sagacity, and with his inferior hand: he never, we believe, has been beaten; and, except in a very few instances of drawn games, has beat the most skillful chess-players in Europe." The Turk was, however, about to encounter its most cunning and perspicacious observer: a young man named Robert Willis.

*

Willis was born in London in 1800, into a medical family. His grandfather Francis Willis was credited with curing George III from his first bout of "madness", which struck in 1788, and was awarded a pension of £1,000 a year by the government. Robert's father, Thomas, was later appointed physician to the king and treated him during a subsequent bout of illness, which is now thought to have been caused by a condition called porphyria. But despite his medical background, Robert Willis, like Maelzel, grew up interested in two things: music and mechanics. In 1819 he patented an improved mechanism for harp pedals, and he was naturally drawn to Maelzel's display of automata when it started to attract attention that year. After seeing the Turk in action, Willis set out to unmask it by publishing a detailed description of its probable mechanism.

Accompanied by his sister, Willis went to see the Turk several times. He first saw it in 1819 at Spring Gardens and noted that the advantage of seeing it in confined surroundings was that it was "more favourable to examination as I was enabled at different times to press close up to the figure while it was playing". Using his umbrella, which he sneaked into the exhibition, he was able to measure the cabinet surreptitiously "with great accuracy" and concluded that it was larger than it seemed, so that there was enough room for a fully grown operator to hide inside it. By December 1820, after several more visits, Willis was ready to publish his findings, though he did so anonymously. His pamphlet, *An Attempt to Analyse the Automaton Chess Player . . . With an Easy Method of Imitating the Movements of that Celebrated Figure,* appeared in early 1821. Willis roundly dismissed "little dwarfs, semi-transparent chess boards, magnetism, or supposing the possibility of the exhibitor's guiding the automaton by wire or piece of catgut so small as not to be perceived by the spectators" in favour of a far simpler explanation involving a hidden operator within the chest.

Willis began his account with a detailed description of the Turk's performance, concentrating in particular on the order in which the doors and drawer were opened and shut, and illustrated with several of his own drawings. The automaton's movements, he noted, "resulting, as they appear to do, from mere mechanism, yet strongly impressed with the distinctive character of an intellectual guidance, have excited the admiration of the curious during a period little short of forty years". But, he added, none of the explanations offered so far had been adequate. He then described three classes of automata: the simple (genuinely autonomous machines), the compound (partially autonomous machines guided by a human agent), and the spurious (machines that appear to be autonomous but are in fact wholly controlled by a human agent). Which class did the Turk fall into?

"The phenomena of the chess player are inconsistent with the effects of mere mechanism," Willis declared, "for however great and surprising the powers of mechanism may be, the movements which spring from it are necessarily limited and uniform. It cannot usurp and exercise the faculties of mind; it cannot be made to vary its operations, so as to meet the ever-varying circumstances of a game of chess." In short, a purely mechanical device could not play chess, so the automaton "must consequently relinquish all claim to be admitted into the first division".

Having dismissed the notion that the Turk was a genuine automaton, Willis moved on to examine the extent of human control over its actions. Was it, as the scientists of Paris had suggested, capable of playing a few moves on its own, with its strategy occasionally guided by its exhibitor?

Was the exhibitor using the influence of a magnet, or tugging a piece of catgut, to tell the automaton which moves to make? Willis noted that building an automaton capable of carrying out even a single chess move would be a fearsome endeavour in itself, but that even if this were possible, "where is the intelligence and the 'Promethean heat' that can animate the automaton and direct its operations? Not only must an intellectual agent be provided, but between such an agent and his deputy, the automaton, a direct communication must be formed and preserved, liable to no interruption, and yet so secret that the penetrating eye of the most inquisitive observer may not be able to detect it."

Whoever was controlling the Turk would, in other words, have to remain in constant communication with it. But, Willis noted with a pointed reference to suggestions of magnets or catgut, "whoever has witnessed the exhibition will have seen that the exhibitor is not confined to a particular spot in the room, but, on the contrary, that he is frequently, during the progress of the game, at some distance from the chest, far beyond the sphere of influence of any of these proposed modes; and if, at such times, the automaton can move a single joint, it is proof decisive that its action springs from another source." This left only one possibility: that the Turk was being operated by someone concealed within the chest.

Before saying how he thought this was done, Willis pointed out a number of oddities about the Turk's presentation.

First, he said, the spectators were briefly shown the machinery in the leftmost third of the cabinet; so when, with a whirr of clockwork, the Turk sprung into action, they naturally assumed that this machinery (now hidden from view behind a locked door) was controlling its movements. But, Willis asked, "where is the evidence that the machinery moves, or that the slightest influence is exerted by it on the arm of the automaton?" If Kempelen had really managed to build such an amazing piece of mechanism, Willis argued, he "would surely be desirous of laying open to view as much of the mechanism of his contrivance, while in actual motion, as he could do". And if the machinery was so secret that it needed to be hidden from view while the Turk was playing, then why was it revealed to the audience in the first place? "The glaring contradiction between eager display on the one hand, and studied concealment on the other, can only be reconciled by considering the exhibition of the mechanism as a mere stratagem, calculated to distract the attention, and mislead the judgement, of the spectators."

Furthermore, Willis pointed out that the order in which the doors of the chest were opened and closed never varied; but why should the order in which the innards of the chest were revealed matter? "This circumstance alone is sufficient to awaken suspicion, for it shows plainly that more is intended by the disclosure than is permitted to meet the eye," he declared. The internal machinery was, Willis concluded, a decoy. To prove this point, he noted that there was no apparent logic to the frequency with which the Turk needed to be wound up. On one occasion, he wrote, the automaton executed sixty-three moves between windings; on other occasions, it needed winding after as few as three moves. Indeed, during one performance Maelzel wound up the Turk and then went back and wound it up again almost immediately, before it had executed a single move.

Willis then provided a detailed description of how an operator might be concealed within the cabinet. First, like Racknitz before him, he suggested that the drawer at the base of the cabinet did not extend all the way to the back. This left a space, he estimated from his measurements, about fourteen inches wide, eight inches high, and four feet long, which was "never exposed to view". But this space alone would not be enough to conceal an adult operator. (This had been the major flaw in Racknitz's argument.) Willis suggested that the machinery in the leftmost third of the cabinet did not extend all the way to the back either, but took up only about one third of the cabinet's depth: "By the crowded and very ingenious disposition of the machinery in this cupboard, the eye is unable to penetrate far beyond the opening, and the spectator is led to conclude that the whole space is occupied with a similar apparatus." The space behind the machinery, he asserted, adjoined the free space behind the draw. Next, Willis suggested that the back of the main compartment was a folding false back. Finally, he claimed that the body of the Turkish figure contained an additional empty space that was never revealed to the spectators.

Taking all of this into account, Willis proposed that the hidden operator entered the cabinet from the left-hand side and sat behind the machinery, with his legs extending forward in the space behind the drawer. Then, by folding down the false back of the main compartment, the operator could make enough space to lean forward while the doors were opened to expose the machinery and illuminate it from behind. Once the back door behind the machinery had been closed, and while the exhibitor opened the drawer at the front, the operator could fold up the main compartment's false back and sit up. At this point, Willis declared, "the success of the experiment may be deemed complete. The secret is no longer exposed to hazard." The front door revealing the machinery could be safely left open, and the front doors opened to reveal the almost-empty main compartment. The cabinet could also be turned around and the doors revealing the innards of the Turkish figure opened.

Once all the doors had been closed, the operator could, suggested Willis, assume the appropriate position to play the game. This involved sliding up into the space within the Turk's torso, from where the operator could see the board through the figure's waistcoat, "as easily as through a veil". And by sliding his left arm through the torso and into the figure's left arm, which Willis assumed was hollow, the operator could then guide the Turk's hand "to any part of the board, and to take up and let go a chessman with no other delicate mechanism than a string communicating with the fingers. His right hand, being in the chest, may serve to keep in motion the contrivance for producing the noise, which is heard during the moves, and to perform the other tricks of moving the head, tapping on the chest, etc." The reason the Turk held a pipe, Willis claimed, was so that its left arm could be suitably shaped to facilitate the insertion of the operator's hand without looking awkward. "The above process is simple, feasible, and effective; showing indisputably that the phenomena may be produced without the aid of machinery, and thereby rendering it probable that the chess player belongs in reality to the third class of automata, and derives its merit solely from the very ingenious mode by which the concealment of a living agent is effected," Willis concluded.

Willis's theory was internally consistent and credible in a way that no previous attempt to explain the Turk's mechanism had been. Within weeks of his pamphlet's publication, Willis was revealed as its author. At the time he was just twenty-one; he would go on to pursue a successful academic career, attending Cambridge University and becoming professor of applied mechanics at the university in 1837. According to one biographer, "his practical knowledge of carpentry, his inventive genius, and his power of lucid exposition made him a most attractive professor, and his lecture-

Engravings from Willis's pamphlet were adapted by Gamaliel Bradford in 1826 to show how Willis thought the Turk worked. He assumed that a hidden operator leaned forward while the machinery was being displayed, and then sat back while the interior of the main compartment was exposed to view. Once the doors were closed, the operator moved up into the Turk's torso and slid his arm into the Turk's left arm to move the pieces.

room was always full." As well as being a gifted lecturer, Willis made a number of contributions to the theory of mechanics, and in later life he also became an authority in the fields of architecture and archaeology.

In putting forward his explanation for the Turk's mechanism, Willis was eager to emphasize that he did not mean to detract from "the real merits of Mr Kempelen; those merits have long since received the stamp of public approbation; indeed, a more than ordinary share of skill and ingenuity must have fallen to his lot, who could imagine and execute such a machine." In contrast to the bluster and fury of Thicknesse, who had previously advocated a similar concealed-operator theory, Willis saw the Turk as a mechanical puzzle to be solved, rather than a fraud to be uncovered. His kinship with Kempelen and Maelzel was further emphasized by his later research into the mechanism of speech, which led him to reconstruct and improve Kempelen's speaking machine.

The notion that no piece of mechanism, however complex, could play chess was the very cornerstone of Willis's argument. This was, he declared, "the province of intellect alone". Yet just as Willis's pamphlet appeared, another young Englishman was coming to exactly the opposite conclusion. He was the man who has since come to embody the notion that a piece of machinery might be capable of logical reasoning: the computing pioneer Charles Babbage.

*

Babbage was still as interested in machines as he had been on his visit to Merlin's workshop two decades earlier. He was seven years older than Willis and had already graduated from Cambridge University. As early as 1812, while he was still a student, Babbage had toyed with the idea that complex mathematical operations might be performed mechanically. "One evening I was sitting in the rooms of the Analytical Society, at Cambridge, my head leaning forward on the table in a kind of dreamy mood, with a table of logarithms lying open beside me," he recalled in his memoirs. "Another member, coming into the room, and seeing me half asleep, called out, 'Well, Babbage, what are you dreaming about?' to which I replied, 'I am thinking that these tables (pointing to the logarithms) might be calculated by machinery.'" But Babbage thought little more of his idea until a few years later, when he went to France to visit the great mathematician Pierre-Simon Laplace.

While in France Babbage saw for the first time the epic mathematical tables that had been compiled under Gaspard de Prony, a French mathematician and engineer. De Prony had been commissioned by the government to draw up a new set of mathematical tables, including the sines of angles to twenty-five decimal places and logarithms of the numbers 1 to 10,000 to nineteen decimal places. Realizing that there were not enough skilled mathematicians in the whole of France to complete the task in a reasonable amount of time, de Prony looked around for a method to speed up the

Charles Babbage

calculations. He soon devised a scheme involving three teams, or divisions, of mathematicians. The first division consisted of half a dozen of the finest mathematicians in France, who worked out the formulas to describe each table; the second division consisted of seven or eight less skilled mathematicians, who boiled down these formulas into simpler numerical problems; and the third division consisted of eighty individuals, most of whom were required only to add or subtract. By breaking down the calculation of each table into a vast number of simple steps, de Prony was able to produce his epic seventeen-volume set of tables in a fraction of the time it would have otherwise taken. In other words, by organizing a group of people so that they performed repetitive actions, like automata, he had been able to automate the process of calculation. This got Babbage thinking about the possibility that calculations could be performed by an automaton using similar principles.

Aside from automata, Babbage was interested in chess, and on March 6, 1819, he saw the Turk play at Spring Gardens. Babbage was evidently intrigued by the Turk, for he acquired a first-edition copy of Windisch's letters, in French, which had been published in 1783. But he found the automaton less impressive as a piece of machinery than as a chess player. "The movement of his hand and arm is not elegant and not so good as many of Merlin's figures," Babbage noted on a piece of paper, which he inserted into this book. "The interior appears large enough for a boy and is lined with green baize. The man who exhibits it stands close to it, sometimes on one, sometimes on the other side. The automaton played very well and had a very excellent game in the opening. He gave check-mate by Philidor's Legacy."

The following year, on February 12, Babbage went to see the automaton again at St. James's Street and challenged it to a game. "Played with the automaton," he wrote. "He gave pawn and the move. Automaton won in about an hour. He played very cautiously – a trap door in the floor of the room was very evident just behind the figure." Babbage was certain the automaton was not a pure machine and was under human control, though he was not sure quite how. But he started to wonder whether a genuine chess-playing machine could, in fact, be built.

In 1821 Babbage and a friend, the astronomer John Herschel, were comparing two sets of astronomical tables, each of which had been calculated independently by different people. The two men's wearisome task was to reconcile the two sets of calculations, comparing each result to ensure that they were identical and therefore likely to be correct. Unfortunately, they were not. "We commenced the tedious process of verification," recalled Babbage. "After a time many discrepancies occurred, and at one point these discrepancies were so numerous that I exclaimed 'I wish to God these calculations had been executed by steam!'" He decided to act. In a spare evening he sketched out a plan of how a calculating machine might work, using no mathematical operations other than repeated addition. Babbage soon came to the conclusion that there was no reason why a mechanical device made of simple parts could not perform complex calculations. He was so excited at the prospect that it made him ill; his doctor advised him to take a holiday and not think about such things, so he went to stay with Herschel near Windsor for a few days. He subsequently drew up a scientific paper in which he announced that he was designing a machine capable of calculating any mathematical tables, including astronomical ones, automatically. This was the genesis of Babbage's first mechanical computer, the Difference Engine.

Babbage struggled in vain for many years and spent the fortune he inherited from his father, along with a vast quantity of government funds, in an unsuccessful attempt to build this machine. Part of the reason why he failed was that halfway through construction, Babbage conceived an even more ambitious machine, the Analytical Engine, which would be capable of far more complex calculations. This caused Babbage to lose interest in the original project. Such was the complexity of this new machine that it was inarguably the earliest ancestor of the modern digital computer: it had direct mechanical equivalents of a modern computer's processor and memory. Babbage even devised a symbolic notation with which to write down programs for it. But following the failure of the Difference Engine, he was unable to raise the funds to build it. Even so, his analysis of the Analytical Engine's theoretical capabilities prefigured many elements of modern computer science. (The Analytical Engine would have relied on punch cards, an idea Babbage borrowed from Joseph Jacquard, whose loom used such cards to represent weaving patterns. Jacquard's loom was, in turn, based on earlier work by Vaucanson.)

In particular, Babbage argued that a suitably powerful mechanical engine would be able to play games of skill such as noughts and crosses, draughts, and even chess. In his memoirs, under the heading "Games of Skill Can Be Played by an Automaton", he noted just how counterintuitive most people thought this idea was. "I endeavoured to ascertain the opinions of persons in every class of life and of all ages, whether they thought it required human reason to play games of skill. The almost constant answer was in the affirmative." But Babbage "soon arrived at a demonstration that every game of skill is susceptible of being played by an automaton". He even sketched out a rough algorithm, or program, for playing board games with movable pieces, including draughts and chess – the first time that anyone had attempted to devise such an algorithm.

Given the near-universal scepticism that an automaton could play a game like chess, Babbage started to consider building a chess-playing machine. It would, he concluded, be far simpler than the Analytical Engine, whose construction he would be able to fund from the proceeds. Proposing to build half a dozen such machines to multiply the income, Babbage looked into previous exhibitions of automata to see how much money they had made. The answer was disappointing; many of the most celebrated automata displayed "were entire failures in a pecuniary point of view". Babbage was by this time in his late middle age and glumly concluded that "even if successful in point of pecuniary profit, it would be too late to avail myself of the money thus acquired to complete the Analytical Engine."

Babbage also considered the question of machine intelligence and concluded that memory and foresight were the defining features of intelligence. In 1834 he bought the silver dancing lady that he had seen in Merlin's workshop, and which had so entranced him as a boy. After restoring it to full working order, he put it on display at his home. In an

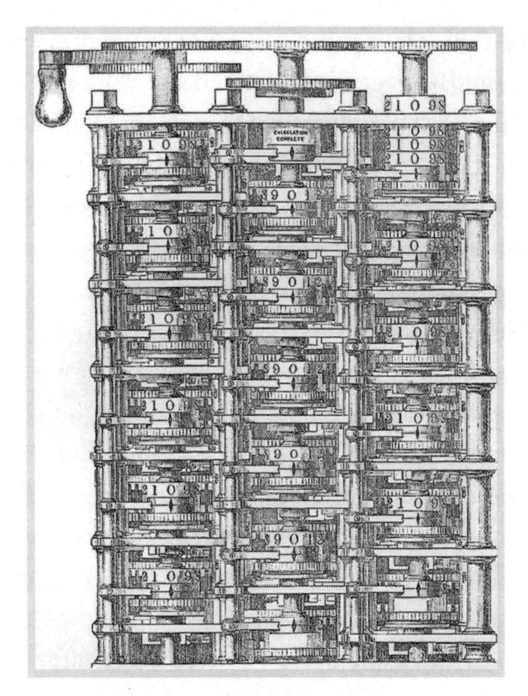

Engraving showing a section of Babbage's Difference Engine.

adjoining room, meanwhile, he displayed an unfinished (but functional) portion of his failed Difference Engine. Babbage liked to draw attention to the difference between these two machines via an elaborate party trick. The dancer, while enchanting, was merely an automaton, since its motions were predictable and preordained; but the numbered dials of the Difference Engine could be programmed to display an ascending series of numbers 0, 1, 2, 3 . . . and so on up to 10,000, at which point the numbers would ascend in increments of three: 10,000, 10,003, 10,006 . . . and so on. The sudden discontinuity would be expected by the programmer, but surprising to the spectator.

The ability to switch from one series of numbers to another suggested that the Difference Engine was more than just a mindless automaton; Babbage contended that this behaviour constituted machine intelligence. The machine appeared to have memory and foresight and could change its behaviour in a way that appeared random, but which was in fact governed by logical rules. Lady Byron, who witnessed such a demonstration with her daughter, Ada Lovelace, wrote to a friend that they had been "to see the thinking machine (for so it seems)". Unlike the new machines of the industrial revolution, which replaced human physical activity, this fragment of the Difference Engine, like the Turk, raised the possibility that machines might eventually be capable of replacing mental activity too.

*

The Turk's appearances in London between 1818 and 1821 thus continued to inspire discussion about the possibility of machine intelligence. Babbage thought that a machine capable of performing logical calculations was theoretically possible; Willis stated categorically that it was not. But despite his careful detective work, Willis had not landed the knock-out punch he had hoped to deliver. His explanation was regarded by most people as just one of the many theories that had appeared over the years, and did little to undermine the Turk's popularity.

In the spring of 1821 Maelzel made the mistake of taking the Turk to Amsterdam, where he ran into Winkel, who was still furious at Maelzel for stealing his design for the metronome. Winkel instituted proceedings against Maelzel for piracy, and the Institute of the Netherlands established a special commission to investigate the case. Maelzel was ultimately forced to admit that he had stolen the design from Winkel, but he then left the country, so the commission's ruling had little effect.

Of more pressing concern to Maelzel were his increasing debts. One of his associates, the Englishman William Lewis, said of Maelzel that "his habits were too expensive. . . . he was generally short of money and at one time owed me fifty pounds." After another season in London, Maelzel moved to Paris and was soon hit with another lawsuit, this time from Eugène de Beauharnais, on the grounds that Maelzel had failed to make payments in accordance with the terms of their agreement concerning the Turk. Eugène's family continued to pursue Maelzel in the French courts even after Eugène's death in 1824, prompting Maelzel to consider selling the Turk to pay off his debts. He contacted his friends among the nobility and at the French court and staged several special exhibitions in an attempt to find a buyer. But nobody would buy the Turk, possibly because it became apparent that it was not truly Maelzel's to sell.

With his debts and legal woes mounting, Maelzel decided to take drastic action. On December 20, 1825, he loaded his collection of automata onto the packet ship *Howard* and set sail for New York, leaving Europe – and, he hoped, his troubles – behind him.

CHAPTER NINE

The Wooden Warrior in America

THE MANHATTAN DEFENCE (D4 D5 C4 E6 NC3 NF6 BG5 NBD7 E3 BB4): A variation of the Queen's Gambit Declined, also known as the American Defence.

The pleasure of a chess combination lies in the feeling that a human mind is behind the game, dominating the inanimate pieces with which the game is carried on, and giving them the breath of life. – Richard Réti

On February 3, 1826, the *Howard* dropped anchor in New York, and Maelzel and his automata disembarked on American soil. Characteristically, Maelzel began the Turk's American campaign in the press, rather than on the stage. Within a few days of his arrival he went to see William Coleman, editor and proprietor of the *New York Evening Post,* who was an ardent chess enthusiast. Coleman published a handsome editorial describing Maelzel's automaton and explained that it would shortly be put on public display. "Nothing of a similar nature has ever been seen in this city that will bear the slightest comparison with it," he wrote.

A few weeks later, on April 13, the Turk was wheeled into the assembly rooms of the National Hotel at 112 Broadway to give its first performance to an expectant crowd. Maelzel went through the routine of displaying the Turk's interior in the usual way and then recruited two volunteers from the audience to play against it. The Turk beat them both easily. The automaton was then wheeled away "amid great and deserved plaudits", according to one eyewitness account. Even though only about 100 people attended this first exhibition, which was a somewhat disappointing turnout, their amazement was such that subsequent performances were sold out and Maelzel started putting on two shows a day, at noon and 8 P.M. At these shows he also displayed the trumpeter and some of his other automata, but it was the mysterious Turk that was the centre of attention. "This wonderful piece of mechanism continues to attract full houses by day and by night," declared one newspaper, two weeks into the Turk's run. "We have repeated our visits with the double view of amusement and, if possible, by close observation, to satisfy ourselves by what process the movements of the figure are regulated and governed, or rather (for after the science and skill of Europe have been exerted in vain, this would be a hopeless task), of determining in our own mind, what are not the causes of his movements. And the result of our observations, thus far, is clearly that the scientific writers have not penetrated the mystery."

The Turk's appearances in New York differed from its previous performances in one important respect. Rather than playing entire games of chess, the Turk played only endgames. In other words, it started from a chess position in which the pieces were placed as though partway through a game. Maelzel had prepared a small book – green, square, about four inches on a side, and morocco-bound – containing seventeen such positions, and the Turk's opponents were invited to choose one of them, as well as the colour they wished to play. Maelzel noted the number of pieces on each side, no doubt to tempt players into choosing the side with the most pieces, rather than with the best position. The Turk, however, always had the first move. So its opponents had to choose the starting position and colour they thought would maximize their chances of winning. Maelzel explained that his decision to restrict the Turk's performances to endgames was because "whole games occupy too much time, and fatigue the attention of those who do not understand the game". But he promised that the Turk would play full games in private, by appointment.

Such was the interest in the Turk that the editor of one newspaper went so far as to apologize to his readers for

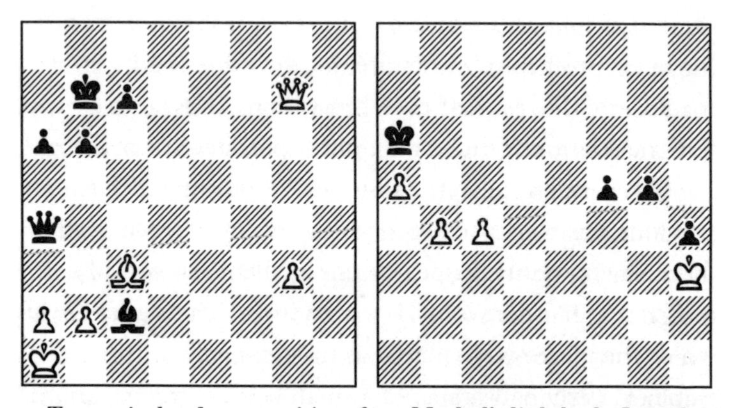

Two typical endgame positions from Maelzel's little book. In each case, victory is guaranteed to whichever side moves first, provided the correct moves are followed. Nevertheless, the Turk lost a game in New York starting from the second position.

"permitting the automaton to occupy so much of his columns". The reason, he explained, was that "persons at a distance can form no idea how much the attention of our citizens is occupied by it." But while Coleman at the *Evening Post* was a staunch ally of Maelzel's, Colonel William Stone, editor of the *Commercial Advertiser,* did not give Maelzel and his automaton such an easy ride. Colonel Stone enthusiastically printed every morsel of information that might help solve the mystery of the automaton's mechanism, and he even tracked down accounts of the automaton's original exhibitions by Kempelen. Maelzel also faced a challenge on another front: within a month of the Turk's first performance, Peale's Museum at 258 Broadway had put on a rival exhibition of "mechanical paradoxes and curiosities made in Philadelphia in imitation of those by Mr. Maelzel" in the hope of profitting from the interest in the Turk. Maelzel had been given his first taste of the combination of intense press scrutiny and fast-moving rival showmen that the Turk would encounter during its travels in America.

Almost immediately, new theories emerged regarding how the Turk worked. A letter in the *National Advocate* noted that Maelzel, who seemed nervous during the Turk's New York shows, often drummed his fingers on the Turk's cabinet. Perhaps, suggested the writer, he was somehow pressing springs or buttons that controlled the automaton's movements. Colonel Stone immediately dismissed the idea on the grounds that Maelzel was usually some distance from the automaton when it made its moves. A rival explanation concerned a striking Frenchwoman who was a member of Maelzel's entourage. She was, it seemed, seen only before or after, but not during, the Turk's performances. Was she somehow controlling the automaton? Maelzel enjoyed scotching this theory by arranging for the woman to sit in the audience during a few performances and even to play against the Turk on one occasion.

All this speculation provided Maelzel with plenty of free publicity and prompted the *Evening Post* to claim that a dozen different theories about the Turk's mechanism had been advanced within a week of its first performance. Perhaps the strangest theory was that Maelzel, who was assumed by many to be the chess-playing brain behind the automaton, indicated to an operator hidden inside the Turk which move to make through the manner in which he picked up and put down the pieces on the Turk's chessboard. (Maelzel had once again adopted the custom of asking the Turk's opponents to sit at a second chessboard at a separate table, while he acted as a messenger between the two chessboards.)

On June 1, after a successful initial run, Maelzel closed his exhibition and announced that he was moving it to Boston. But he subsequently reopened his show and did not leave New York until July 5. The reason for this delay is unclear. Perhaps Maelzel wanted to remain in the city for the fiftieth anniversary of the Declaration of Independence on

July 4; he may also have been resolving his legal difficulties with the estate of Eugène de Beauharnais during this period. Representatives of Eugène's family had pursued Maelzel to New York, and only by paying off the outstanding amount owed – variously reported as $800 or $4,000 – was he able to settle the suit once and for all.

Maelzel was anxious to leave New York because he was coming under increasing pressure from one of New York's leading chess players, a man known as Greco, who wanted to play the Turk in a private match. Maelzel had promised that the Turk was available for private games by appointment, but he would not agree to let Greco or his friends play and fobbed them off with a series of implausible excuses. Greco assumed that Maelzel was afraid that the Turk would lose, and that this might cause the public to lose interest in the automaton, which had suffered only two defeats while in New York. But despite his constant pestering, Greco failed to get his way by the time Maelzel left for Boston.

*

In Boston Maelzel set up his exhibition at the Julien Hall, on the corner of Milk and Congress Streets. The Turk gave its first performance on the evening of September 13, and after that the automaton gave two performances a day, playing endgames as it had in New York. Within a few days it suffered a defeat. According to the *Boston Centinel*, which was unaware that the Turk had twice been defeated in New

York, this clearly demonstrated the superiority of Boston's chess players: "On Monday the grave and skillful chess-player found a conqueror in a Bostonian, in one of his most favorite end of games, and was compelled to succumb, we believe for the first time since his arrival in America."

This infuriated the chess players of New York, where the *Centinel*'s article was reprinted in the *New-York American*, which reminded its readers that the Turk had in fact been defeated twice during its New York run. But before long the Turk had suffered a total of three defeats in Boston, although on two of these occasions Maelzel had given the Turk's opponent the first move. So the *Centinel* was able to respond smugly: "The truth is that the automaton has been conquered in Boston by three gentlemen separately."

With the honour of New York's chess players now at stake, Greco wrote a furious letter to the *New-York American*. He pointed out that the endgames did not represent a true test of the automaton's abilities, and that "the only fair test of skill would be full games". Greco explained that he had tried and failed to get Maelzel to agree to a full game. "Although Mr. Maelzel had advertised amateurs might have this privilege," he complained, "he always evaded the application and it was impossible to encounter his 'wooden warrior' in a complete game. Believing as I do, however erroneously, that at least two people in New York can play with a degree of skill inferior to no champion whatsoever, either American or European, I wish to challenge the automaton chess-player to play a match of three full games on its return to the city. The match to be contested for love or money, as Mr. Maelzel pleases. If he accepts this defiance, the American public will be pleased to ascertain whether our countrymen are equal or not to foreigners in any specified particular – even a knowledge of the game of chess."

If the automaton refused to accept the challenge, Greco concluded, "let us hear no more of its vaunted superiority to the world as a chess player, but content ourselves with admiring it as a most admirable instance of mechanical ingenuity." Greco had thus raised the stakes; he was claiming that New York's chess masters not only were better players than their rivals in Boston but were the finest players in the world. Thrilled at the publicity this spat was generating, Maelzel gleefully doubled the entrance fee to his Boston exhibition to one dollar. And on October 13, in a shrewdly calculated snub aimed directly at Greco, he also announced that in his opinion, the Boston players were every bit as good as the New Yorkers, and that the automaton would accordingly play its first week of full games in Boston.

The automaton played its first full game starting at noon on October 16, and its performance was dazzling. For the next few days it easily defeated a succession of Boston's most skilled players. But at the end of the week the Turk was challenged by Benjamin D. Green, a young doctor, who went on to win the game, to the delight of his fellow Bostonians. For years afterward Green was better known as "the man who beat the automaton" than he was as a doctor.

Maelzel left Boston on October 28 and set out for Philadelphia, stopping in New York on the way in order to deal with Greco. He announced that, with the backing of the chess players of Boston, he would be happy to accept Greco's challenge, with the stakes to be set at between $1,000 and $5,000 at Greco's discretion. This seems to have rattled Greco, for rather than agree to a public match, he arranged for the two chess players he had referred to in his letter to visit Maelzel privately. The Turk was, at this point, still packed into its crates, and rather than unpack it Maelzel asked the two men whether they would consent to play against his assistant and secretary, a young Frenchman named William Schlumberger. Greco had, after all, claimed that his two champions were capable of defeating not just the automaton but any American or European player too. If the Americans proved to be strong players, Maelzel evidently planned to make a private deal with them to avoid a public defeat for the automaton.

The two men agreed to play Schlumberger, and he swiftly defeated both of them, forcing Greco into a humiliating retreat. "Since my former communication," Greco wrote to the *New-York American,* "I am very sorry to state that both the American chess players on whose skill I had relied so arrogantly have been beaten with ease by a foreigner.

I must therefore back out from my challenge, as better men have done before me, and subscribe to the automaton's superiority without a trial."

*

As Maelzel prepared to open his exhibition in Philadelphia, a pamphlet about his automaton written by Dr. Gamaliel Bradford was published in Boston. *The History and Analysis of the Supposed Automaton Chess Player of M. de Kempelen* presented a comprehensive roundup of the various theories about the Turk's mechanism, with a few unusual new ideas thrown in for good measure, though Bradford did not commit himself firmly to any particular theory. "The arrival of this celebrated Androides was anticipated in this country with much interest, and abundantly answered the expectations of the curious," he began. "Many theories have been framed to account for its operation, but all are attended with serious difficulties."

Bradford outlined Racknitz's theory that a hidden dwarf operated the machine, but rejected the idea of magnets under the chessboard to reveal which piece had been moved in favor of the transparent chessboard favoured by Decremps. Then he considered the possibility that the automaton was entirely controlled by Maelzel, who told it how to move by pressing the appropriate squares of its chessboard, which might in fact be "moveable cubes, connected with appropriate springs". The idea was that Maelzel surreptitiously pressed these squares while standing by the Turk to move the chessmen on behalf of the Turk's opponent. But Bradford rejected this theory on the grounds that the delay between the opponent's move and the Turk's response varied from a few seconds to three minutes. If Maelzel was actually telling the automaton what move to make by pressing buttons, Bradford reasoned, this delay would always be the same length. Furthermore, Maelzel did not always touch the machine, and he touched it in different places.

This raised another intriguing possibility. Perhaps, wrote Bradford, there were a multitude of buttons hidden around the machine that prompted the Turk to make different moves. This would explain why Maelzel touched the machine in different places at different times. The problem with this idea, however, was that there would have to have been at least 1,600 buttons to accommodate all the possible moves that can occur during a game of chess. (There are sixty-four squares on a chessboard, so it might seem that $64 \times 63 = 4{,}032$ buttons would be needed to encompass all possible pairs of squares, but in practice most of these pairs represent impossible moves.) And even 1,600 buttons would have been impractical. "Some have supposed that they were arranged like the keys of an organ, along the moulding of one end of the box, but this end being two feet long, it would follow that there must be 66 to an inch, and that they must present a surface too narrow to be separately touched," Bradford noted.

He then moved on to consider the possibility that Maelzel was controlling the machine using magnets hidden in his pockets, but dismissed this idea because Maelzel was often several feet away from the automaton. Might an assistant, hidden behind the curtain or beneath the floor, be responsible for operating the machine? "At such a distance magnetism is out of the question," wrote Bradford. "Communication through the floor is evidently impossible, and that by strings or wires from the roof equally so. The assistant, therefore, if anywhere, is in the box." Bradford then outlined Willis's explanation of how an operator might be concealed within the box, though he disagreed with Willis's suggestion that the operator viewed the game through an aperture in the Turkish figure's chest. Instead, Bradford suggested an elaborate alternative theory: that of "a common camera obscura apparatus, of which the lens is in one of the eyes of the automaton, the mirror being situated within the head, at such an angle as to reflect the rays of light towards a plate of ground glass placed in the back of the box, and near the occupant". This would have given the operator a Turk's-eye view of the board.

All of these theories, Bradford concluded, merely added to the Turk's mystique. "There is much hear-say evidence abroad, which goes either to confirm or refute the explanation above given," he wrote. "Whether the secret will be completely exposed in our time is uncertain, but whether it shall turn out to be a mere machine, directed by springs and wheel-work set in motion by the exhibitor, or an assistant at a still greater distance, or a mere puppet moved by a player within, who has for half a century eluded the observation of thousands of eagerly watchful spectators, it must be admitted to be one of the ingenious and completely successful contrivances which has ever been offered to the public; instead of satisfying, it seems continually to excite curiosity, and the more one goes to see it, the more desirous he becomes to visit it again."

*

Maelzel opened his show in Philadelphia on December 26, 1826. He had spent several weeks getting ready, during which time he had taken an extended lease on a building on Fifth Street and spent a considerable amount of money refitting it. A team of workmen put in a new staircase, decorated the exhibition room, and converted the remaining space into a workshop and private quarters for Maelzel and his associates. Maelzel decided within a few days of his arrival in Philadelphia that he would make the city his base, and quickly established contacts in the local business, music, and scientific communities. The building where he lived and held his exhibitions became generally known as Maelzel's Hall.

As usual, the Turk appeared twice a day and played mainly endgames, interspersed with a few full games. Between December 26 and March 20, when its run in

Philadelphia ended, it lost one endgame and one full game, played against a woman named Mrs. Fisher. Details of the game, which were published in the *Philadelphia Gazette,* reveal that the Turk played unusually badly and resigned after thirty-nine moves; it has been suggested that Maelzel deliberately ensured that the automaton would lose the game in order to flatter Mrs. Fisher. Another of the Turk's full games, which lasted five hours and was played in three sittings, ended in a draw.

From his new base in Philadelphia, Maelzel decided to visit Baltimore, and he opened his exhibition there on April 30 at the Fountain Inn on Light Street. The Turk's appearances were preceded by performances from the automaton trumpeter, "the amusing little Bass Fiddler" (another musical automaton), and "the automaton slack rope dancers", a set of mechanical trapeze artists which had by this time become one of Maelzel's favourite automata. As in Philadelphia, the Turk played mainly endgames and rarely lost. On one occasion it was, however, defeated almost accidentally by Dr. Joshua Cohen, an acquaintance of Maelzel's who was invited to play an endgame against the automaton after attending several of its performances.

Cohen stepped forward reluctantly and was shown a starting position in the little green book by Maelzel, who asked him which colour he wished to play and offered him the concession of moving first. Cohen deliberately chose black, the colour with fewer chessmen and without a queen,

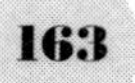

since he assumed that the starting position was a trick, and that Black actually had the stronger position. Unwilling to embarrass Maelzel, he made his first few moves without thinking too hard, expecting to be checkmated almost immediately; but he then found himself in a rather strong position and began to apply himself to the game. Before long, to Cohen's embarrassment, it became clear that victory was in his grasp. Once it became clear that the Turk was going to lose, Maelzel suggested that Cohen make an illegal move, to demonstrate the automaton's response. Cohen did so, and the Turk duly moved the offending piece back to its original position and then made its own move. The audience was delighted, except for the small minority who realized that Cohen had thrown away a certain victory, and cried "But how about the game?" to no avail. The game was reported in the following day's newspaper as a victory for Cohen, who felt so bad about his inadvertent win that he even apologized to Maelzel.

Another unusual defeat came on May 23, 1827, when the Turk played against Charles Carroll, the last surviving signatory of the Declaration of Independence. The automaton played far below its usual ability but came within one move of victory nevertheless, at which point Maelzel claimed that he needed to adjust the Turk's mechanism. Taking a candle, he knelt down and opened one of the Turk's rear doors. Once Maelzel's "adjustment" was complete, the Turk made an idiotic move and managed to lose the game. Carroll

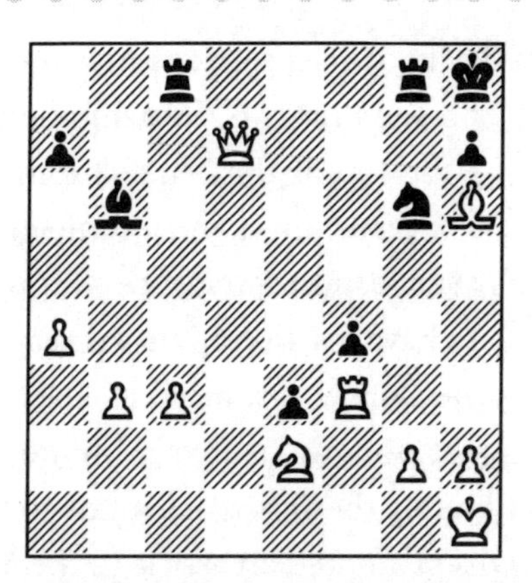

The starting position for the endgame played by Dr. Joshua Cohen against the Turk in Baltimore in May 1827. Cohen correctly assumed that even though White has more chessmen and the advantage of a queen, Black's is the stronger position. In fact, Black can win in seven moves.

must have realized what was going on; "I think you have favoured me in this game," he remarked to Maelzel afterward. The *New York Mirror* subsequently reported that Carroll, aged eighty-nine, "beat his Turkish majesty in a game of chess in the city of Baltimore, to the great delight of a crowded audience."

These occasional defeats had little effect on the public's enthusiasm for the automaton. But an article that appeared in the *Baltimore Gazette* on June 1 appeared to spell disaster for Maelzel. Its headline was "The Chess-player Discovered", and it reported the account of two boys who were said to have found out the Turk's secret. According to the boys, on a hot day in the last week of May they had climbed onto the roof of a shed next to the exhibition hall and saw Maelzel opening the top of the Turk's cabinet after one of its exhibitions – whereupon a man climbed out. Maelzel brushed off this suggestion, saying he was used to people falsely claiming to have discovered the secret of the automaton. On June 5 the *Gazette* distanced itself from its original report, noting that nobody had corroborated the boys' report, and that one of them had asked for money in return for his story. This prompted an article in the *National Intelligencer* in Washington, D.C. – one of the most respected newspapers in the country – which claimed that Maelzel himself was behind the story. "The tale of a discovery was but a clever device of the proprietor to keep alive the interest of the community in his exhibition," the newspaper declared, mocking the *Gazette* for falling for such an obvious publicity stunt. As a result, no other newspapers picked up the story, and it was swiftly forgotten.

Maelzel had, in any case, decided to close his Baltimore exhibition for the summer on June 2. A shipment had just arrived from Europe of some of his other exhibition pieces, including the Conflagration of Moscow. Maelzel announced that his exhibition would soon reopen, with this new exhibit on display in addition to the Turk and his other automata. But it would be some time before Maelzel opened his doors in Baltimore again, for he had to deal with a new and unexpected threat to the Turk. In New York an impostor had appeared: a rival chess automaton.

*

On April 22, 1827, the American Chess Player made its first appearance at the American Museum in New York. As this automaton's name suggested, it was an American-built imitation of the Turk, and had been constructed by two brothers named Walker in conjunction with a mechanic called Bennet. Maelzel had been made aware of the Walkers' intention to build an imitation Turk a few months earlier. "You Americans are a very singular people," he later recalled to one of his friends. "I went with my automaton all over my own country – the Germans wondered and said nothing. In France, they exclaimed, *Magnifique! Merveilleux! Superbe!* The English set themselves to prove – one that it could be, and another that it could not be, a mere mechanism acting without a man inside. But I had not been long in your country, before a Yankee came to see me and said 'Mr Maelzel, would you like another thing like that? I can make you one for five hundred dollars.' I laughed at his proposition. A few months afterwards, the same Yankee came to see me again, and this time he said 'Mr Maelzel, would you like to buy another thing like that? I have one ready made for you.'"

Maelzel had brushed off the offer, but now that the Walkers had put their automaton on display, he could ignore it no longer. Worse, newspaper reports were claiming that the American Chess Player was as impressive as the Turk, even though it was not as strong a player. Maelzel wrote to his friend Coleman to ask him what he thought. To his alarm, Coleman replied that he thought the impostor was the better automaton of the two. So Maelzel hurried to New York, and he and Coleman went to see the American Chess Player for themselves.

The automaton was not as finely decorated as the Turk, but it was slightly smaller and appeared to contain more machinery. Like the Turk, it had a speaking machine inside it that enabled it to say "Check" when appropriate. But it was not a very good player, and the Walkers were not very accomplished showmen. After the show, Maelzel offered the brothers $1,000 if they would withdraw their automaton from public display, but they refused. Maelzel was unwilling to offer them a larger bribe and concluded that the rival automaton posed little threat, because it was simply not very good at playing chess and lacked the aura of near-invincibility (not to mention the historical credentials) of his own automaton. But the appearance of the rival automaton seems to have alerted Maelzel to the danger of relying too heavily on the popularity and uniqueness of a single exhibit. He returned to Baltimore and reopened his exhibition in October, with the Conflagration of Moscow as the main attraction; a footnote on the poster promoting his exhibition noted that "the automaton chess player will be exhibited only to private parties, on application to Mr Maelzel."

Maelzel returned to Philadelphia in late November and was annoyed to discover that the American Chess Player was now being displayed on his home turf. He reopened his exhibition there in January 1828, again with the emphasis on the Conflagration of Moscow, and then moved on to Boston in May. On the way he stopped in New York to warn the public about the rival chess player. "The impostor," Maelzel's advertisement in the *Evening Post* pointed out, "was not the automaton he had exhibited in New York two years ago, nor had it any real pretentions to the skill and power of that celebrated chess player." A few days later the rival automaton's new proprietor responded that his chess player was "at least the equal" of Maelzel's automaton, "if not superior". But its fortunes were in decline, and a few weeks later an advertisement appeared offering the American Chess Player for sale. It has been claimed that Maelzel himself purchased the automaton for $5,000 and then destroyed it; but it seems more likely that the American Chess Player was bought by Eugene Robertson, a showman and balloonist, who took it to Mexico. In any case, the impostor was never heard of again.

Yet just as one rival automaton had been disposed of, another appeared: the Automaton Whist Player, which went on display in New York in May 1828. It did not play chess, but Maelzel still considered it a threat to the Turk and arranged to buy it from its owner. This new automaton became part of Maelzel's exhibition for a while but was never very popular. Its acquisition was, however, part of a broader plan initiated by Maelzel in the summer of 1828 to expand and diversify his exhibition. In August he sold several of his exhibits, including the Conflagration of Moscow and the slack-rope dancers, to a group of Boston businessmen for $6,000. He agreed to let them display these items under his name, and Maezel's Exhibition opened in Boston in September and then went on tour, visiting Providence, Baltimore, Richmond, and New York. While this tour was under way, Maelzel put the Turk into storage in the care of a trusted associate and returned to Europe to search for new automata for his exhibition.

Maelzel arrived back in New York in April 1829 with several new exhibits, including the Mechanical Theatre, specifically intended for children, a display of automaton horse riders called the Grand Tournament, and a diorama of the interior of Rheims Cathedral. He entered into a partnership with the Boston company, whose exhibition was not doing very well without his theatrical flair, and brought the Turk out of storage. The resulting exhibition, amassed in New York, was so large that Maelzel divided it into two separate shows, one at Tammany Hall and the other at 223 Broadway. The New York shows ran until May 1830, and after that Maelzel continued to tour on his regular circuit of Philadelphia, New York, and Boston until 1834.

Maelzel's
EXHIBITION,
MASONIC HALL.

☞ PERFORMANCE EVERY EVENING. ☜

ON SATURDAY, MAY 17 1834
There will be two Exhibitions, one commencing at M. the
at the usual time.—Doors open half an hour previ

Doors open at half-past 7 o'clock. Performance to commence at 8 preci

PART FIRST.
THE ORIGINAL AND CELEBRATED
AUTOMATON
CHESS PLAYER.
Invented by DE KEMPELIN, Improved by J. MAELZEL.

The Chess Player has withstood the first players of Europe and America, and excites universal admiration. He moves his head, eyes, lips, and hands, with the greatest facility, and distinctly pronounces the word *"Echec,"* (the French word signifying *"Check"*) when necessary. If a miss-move is made, he perceives and rectifies it.

THE
Automaton Trumpeter.

The Trumpeter is of a full size, and dressed in the uniform of the French Lancers. The pieces executed by this Automaton are performed with a distinctness and precision unattainable by the best living performers; the measurement of time being, from the nature of the mechanism, absolutely perfect. In double-tongueing, his superiority is particularly manifested, not only in the clearness of the tones, but also in the number of the notes which are sounded. All the sounds are actually produced in the Trumpet, there being no pipes whatever within the figure. The pieces he plays were written expressly for him by the first composers. He will perform on each evening, two favorite pieces. 1st, the French or Austrian Cavalry Manoeuvres. 2d, A March, with an accompaniment.

THE
MECHANICAL THEATRE,
Purposely introduced for the gratification of Juvenile Visiters.
IT CONSISTS OF THE FOLLOWING PIECES;

1 The *Amusing Little Bass Fiddler.*
2 The *French Oyster Woman*—who bows to the Company, and performs the duties of her station, by opening and presenting her Oysters to the audience.
3 The *Old French Gentleman,* of the ancient Regime, who drinks the health of the Company with great glee.
4 The *Chinese Dancer,* accompanying the Music with his Tamborine.
5 The *Little Troubadour,* playing on several instruments.
6 *Punchinello* will go through his comical attitudes in imitation of the celebrated Mazurier.

Part of a handbill advertising Maelzel's Exhibition at the Masonic Hall in Philadelphia May 17, 1834.

P. T. Barnum

A promotional poster for an 1834 show in Philadelphia gives a typical lineup: the automaton trumpeter, followed by the Mechanical Theatre, the slack-rope dancers, the Grand Tournament, the diorama of Rheims Cathedral, a musical automaton called the Melodium, and the Turk as the grand finale. Silas Weir Mitchell, a Philadelphia doctor who attended Maelzel's show as a child, later recalled that "the Turk, with his oriental silence and rolling eyes, would haunt your nightly visions for many an evening thereafter."

In August 1834 Maelzel journeyed south to Richmond, Virginia, and continued on to Charleston, South Carolina, where the Turk was put on display in late November. The following year Maelzel was back in Boston, where he met a young showman who later wrote in his memoirs that he saw Maelzel as "the great father of caterers for public amusement, and was pleased with his assurance that I would certainly make a successful showman". The young man in question was P. T. Barnum, who recalled that Maelzel gave him one piece of advice: to exploit press coverage of his exhibitions to the fullest. "I see that you understand the value of the press, and that is the great thing," Maelzel told Barnum. "Nothing helps the showman like the types and the ink."

In late 1835 Maelzel had two exhibitions running: one in Washington, D.C., and the other, including the Turk, in Richmond. During the automaton's Richmond run Maelzel encountered another talented young man who, like Barnum, was at the beginning of what would prove to be a highly successful career. This second young man was a journalist, not a showman; he wrote what was to become by far the most famous and widely available account of the Turk, and in the process helped define a new form of literature. His name was Edgar Allan Poe.

CHAPTER TEN

Endgame

ENDGAME: The final phase of a game, when there are few chessmen left on the board.

As the strong man exults in his physical ability, delighting in such exercises as call his muscles into action, so glories the analyst in that moral activity which disentangles. He derives pleasure from even the most trivial occupations bringing his talent into play. He is fond of enigmas, of conundrums, hieroglyphs; exhibiting in his solutions of each a degree of acumen which appears to the ordinary apprehension preternatural.

– Edgar Allan Poe, "The Murders in the Rue Morgue"

At the time of his encounter with the Turk in 1835, Edgar Allan Poe was twenty-six years old and was living in Richmond, Virginia. He had launched his journalistic career two years earlier, when he won a prize of $100 in a prose-writing contest organized by a Baltimore periodical. This enabled him to get a job on the *Southern Literary Messenger,* a Richmond magazine, though he was fired a few months later because of his tendency to drink too much. But in the autumn of 1835 Poe was offered the post of editor at the magazine. Under his stewardship, the *Messenger*'s circulation would increase from 500 to 3,500 over the next two years, during which time Poe wrote eighty-five reviews, six poems, three stories, and four essays. One of these essays was inspired by the Turk's visit to Richmond, where Poe saw it play in December 1835.

Titled "Maelzel's Chess-Player", the essay appeared in the *Messenger* in April 1836. It was an analysis of the Turk's operation that drew heavily on the description provided by David Brewster in his popular book *Letters on Natural Magic.* Brewster's work was, in turn, heavily reliant on that of Robert Willis. But Poe also added several observations and deductions of his own, "taken during frequent visits to the exhibition of Maelzel". And he presented his conclusions in a format that prefigured his later mystery and detective stories.

"Perhaps no exhibition of the kind has ever elicited so general attention as the chess-player of Maelzel," Poe began. "Wherever seen, it has been an object of intense curiosity to all persons who think. Yet the question of its *modus operandi* is still undetermined. Nothing has been written on this topic which can be considered as decisive." After a brief summary of the history of automata, touching on the work of Vaucanson and others, Poe explicitly compared the Turk to Babbage's calculating engines. The Turk, he noted, had to respond to its opponent's moves, whereas Babbage's machine, while impressive, merely performed a series of mathematical operations that were entirely predetermined by its initial configuration. He concluded that "if we choose to call the former a pure machine we must be prepared to admit that it is, beyond all comparison, the most wonderful of the inventions of mankind." In reality, he declared, "it is quite certain that the operations of the automaton are regulated by mind, and by nothing else. . . . the only question is of the manner by which human agency is brought to bear."

Poe then gave a brief account of the automaton's history and described the manner of its presentation, detailing the order in which the doors were opened and shut, and so on.

He explained the use of a second chessboard and noted that this required Maelzel to pass to and fro between the automaton and the table every time a move was made. He observed that Maelzel "frequently goes in the rear of the figure to remove the chess-men which it has taken, and which it deposits when taken, on the box to the left (to its own left) of the board". Poe also detailed some of Maelzel's strange behaviour during the Turk's performances: "When the automaton hesitates in relation to its move, the exhibitor is occasionally seen to place himself near its right side, and to lay his hand now and then, in a careless manner, upon the box. He has also a peculiar shuffle with his feet, calculated to induce the suspicion of collusion with the machine in minds which are more cunning than sagacious. These peculiarities are, no doubt, mere mannerisms of Mr Maelzel, or, if he is aware of them at all, he puts them in practice with a view of exciting in the spectators a false idea of the pure mechanism in the automaton." The illustrations in Poe's article reveal that the Turk was now playing with a feather in its turban.

Having described the Turk's performance, Poe turned to the various theories that had been proposed to explain how it worked. He dismissed the theories of the hidden dwarf or small child and accused Willis of following "a course of reasoning exceedingly unphilosophical". Poe's objection was that Willis simply outlined a way in which the Turk might have worked, but did not provide enough observational evidence to back up his claim that his solution was the correct one: "We object to it as a mere theory assumed in the first place, and to which circumstances are afterward made to adapt themselves." In contrast, wrote Poe, "in attempting ourselves an explanation of the automaton, we will, in the first place, endeavour to show how its operations are effected, and afterward describe, as briefly as possible, the nature of the observations from which we have deduced our result."

Accordingly, Poe described how he thought the Turk worked, largely parroting Willis's explanation. The automaton was, he declared, controlled by a hidden operator who changed his position and adjusted a series of movable partitions inside the cabinet, in order to stay one step ahead of

Edgar Allan Poe

the audience as various parts of the cabinet's interior were revealed. The operator then, he claimed, moved up into the Turk's body to look out over the chessboard and operate the automaton's arm. "Our result is founded on the following observations," Poe declared, and went on to present a list of seventeen pieces of supporting evidence.

The Turk, he noted, took longer to move in some cases than in others, which suggested that the automaton was operated by a hidden human agent. Poe also observed that the automaton did not always win, and asserted that it would always have won if it had been a pure machine, on the basis that it would be only slightly more difficult to build a machine capable of winning all games than a machine capable of winning some games. The Turk only rolled its eyes and shook its head (in apparent exasperation at the incompetence of its opponent) during easy games, Poe pointed out; this was also suggestive of a hidden operator, who would not have had time for such tomfoolery during a more difficult game. Poe also thought that the automaton's jerky movements were suspicious, because other automata were so smooth and lifelike; he suspected this was a ploy to make the Turk seem more like a machine, when it was really nothing of the sort.

Poe went on to propose that the automaton's cabinet was larger than it seemed, and lined with cloth to deaden the sounds made by the hidden operator. The two candelabras placed on the top of the cabinet during the performance were, he suggested, necessary to provide enough light for the operator to see through the gauze of the Turkish figure's chest; the Turk's opponent was granted only a single candle. Poe noted that Maelzel's secretary, Schlumberger, was never seen during the performances, and that on one occasion when Schlumberger was unwell, exhibitions of the Turk were suspended. He also thought that the fact that the Turk played with its left arm was significant; he suggested that this was to accommodate the hidden operator more easily. Individually, not all of these pieces of evidence are terribly convincing, but collectively they add up to a persuasive argument. "We do not believe that any reasonable objections can be urged against this solution of the automaton chess-player," he concluded.

Poe's explanation thus has the structure of a short detective story of the sort that he was later to become famous for writing. After a brief prologue, the mystery is introduced, and the evidence, replete with numerous red herrings, is laid out. Once all other investigators have professed themselves baffled, the detective makes a number of observations whose significance is not immediately apparent. He then appears to leap directly to the solution; only at the end of the story does he explain how several steps of reasoning, confirmed by his observations, enabled him to solve the mystery.

A good example of this format is Poe's later story "The Murders in the Rue Morgue", which starts with newspaper accounts detailing two grisly murders, along with the reports of witnesses. The amateur detective, Dupin, visits the scene of the crime and makes a few crucial observations that are unnoticed by the story's narrator. Dupin then announces that he has solved the mystery, and that he expects the murderer's accomplice to arrive at his home shortly in response to a newspaper advertisement that he has just placed. While they await the accomplice's arrival, Dupin explains how he arrived at his hypothesis and how his observations subsequently confirmed it. The accomplice then arrives and confesses all, confirming Dupin's explanation.

Seen in this light, Willis's explanation of the Turk was indeed defective, because Willis never explained the reasoning by which he arrived at his conclusion, and why he believed it to be correct; he merely outlined a way in which the Turk might have worked. Poe, in contrast, justified his explanation on the basis of observation (though some of his justifications were, admittedly, more pertinent than others). All that was missing was an admission from Maelzel that Poe's explanation was correct, and that he was guilty as charged of presenting a man in a box as a chess-playing automaton. Poe wrote that "when the question is demanded explicitly of Maelzel: 'Is the automaton a pure machine or not?' his reply is invariably the same: 'I will say nothing about it.'" He even went so far as to challenge Maelzel to respond to his essay, according to one newspaper article, which noted that "the essay on the automaton cannot be answered, and we have heard the editor challenges a reply from Maelzel himself, or from any other source whatever." But Maelzel did not respond. By the time Poe's essay appeared in print, Maelzel and the Turk were back in Philadelphia, where the automaton went on display for two months beginning in late April.

The format of Poe's analysis of the Turk is widely regarded as the prototype for his later mystery stories, and the essay functions as a trial run for the didactic literary voice of Dupin, whose character was, Poe later admitted, modelled on himself. Writing in 1926, J. W. Krutch, one of Poe's many biographers, described "Maelzel's Chess-Player" as "the first extended example of the author's skill in what he called 'ratiocination'", which is the word Poe liked to use to describe the process of logical reasoning. Another biographer, Hervey Allen, declared that "it was the first of Poe's works in which he emerged as the unerring, abstract reasoner, and foreshadowed the method he followed later in his detective stories such as 'The Murders in the Rue Morgue' – a method which has been embalmed in the triumphs of Sherlock Holmes." In reality, Poe's essay was not quite the flawless piece of detective work that his admirers have suggested. Poe relied more heavily on Brewster and Willis than is often acknowledged, and some of his arguments, while seeming logical, are bogus (such as his contention that a machine that could win every game would be only slightly harder to build than one that lost occasionally). And most important, his explanation of the Turk's mechanism was actually incorrect.

Even so, unlike previous attempted exposés of the Turk, Poe's essay caused quite a stir. The *Norfolk Herald* commented that "the piece has excited great attention", and newspapers up and down the East Coast reprinted it and quoted from it. True or not, Poe's explanation was certainly convincing, and its popularity convinced Maelzel that the time had come, once again, to pack up his exhibition and move on.

*

In the autumn of 1836 Maelzel took his exhibition west to Pittsburgh and then travelled by steamboat along the Ohio River, stopping in Cincinnati and Louisville. From there he traveled down the Mississippi to New Orleans, arriving just after Christmas, where his exhibition ran until the end of February 1837. He then took a ship to Havana, where he staged a short but successful exhibition. In addition to the Turk, Maelzel's show included the automaton trumpeter, the Melodium, the slack-rope dancers, the Mechanical Theatre, and a new diorama called the Pyric Fires. But where, asked the spectators, was the famous Conflagration of Moscow? The answer was that Maelzel had sold it, so he resolved to build a bigger, better version and to bring it to Havana the following year, after which he would embark on a tour of South America.

Maelzel returned by ship to Philadelphia, where he discovered that the Turk had been the subject of yet another exposé. The source was an article that had appeared in *Le Palamède,* a French chess journal, in 1836 and was reprinted in translated form in the *National Gazette* on February 6, 1837. Its author, Mathieu-Jean-Baptiste de Tournay, essentially followed Racknitz's explanation of the Turk's mechanism – namely, that it was controlled by a hidden operator who followed the action on the chessboard via a system of magnets. Like Poe, de Tournay insisted that he had at last revealed the automaton's secret; also like Poe, he was ignored by Maelzel, who had become engrossed in the construction of his ambitious new diorama. It was an expensive undertaking that lasted the whole summer and autumn, during which time Maelzel did not mount an exhibition in Philadelphia. Such was Maelzel's attention to detail that the construction of the diorama overran both the time and funds available, and he had to borrow money from his friend John Ohl, a Philadelphia businessman, to pay the carpenters and painters he had hired to build it. Eventually, on November 9, Maelzel set out for Havana again on board the *Lancet,* a small ship that belonged to Ohl.

The exhibition was ready to open by the end of December, and Maelzel's intention was to keep it open throughout the carnival season, which ran from Christmas to the end of February. The new Conflagration of Moscow was particularly well received, prompting Maelzel to extend his exhibition into March 1838. But this proved to be a mistake. The season of Lent had begun, and attendance fell off dramatically. But worse was to come. In April Maelzel's secretary and close friend Schlumberger caught yellow fever and died, at which point the other members of Maelzel's company deserted him. Maelzel, now sixty-five years old, suddenly found himself alone in a foreign land, heavily in debt, with only his automata for company.

Eventually Francisco Alvarez, one of Ohl's business associates in Havana, took pity on Maelzel and loaned him some money. With his planned tour of South America now out of the question, Maelzel halfheartedly wrote to an associate in Philadelphia to ask him to arrange a new exhibition hall in preparation for his return. On July 14 he set sail from Havana on board the *Otis,* another of Ohl's ships. But Maelzel was by this time a broken man. Captain Nobre of the *Otis,* who had seen Maelzel in Havana a few months earlier, was shocked by the sudden decline in his passenger's health. On the first evening of the voyage, seeing Maelzel sitting on the deck clutching a travelling chessboard, the captain tried to cheer him up with a game of chess, even though he was a weak player. Maelzel beat him easily in the first game, but became uncharacteristically bad-tempered and withdrawn during the second, and lost. It was a sad end to the last game Maelzel would ever play.

That night Maelzel retired to his cabin, where he asked the steward to place a case of wine, which he had brought on board with him, within reach of his berth. For six days he sulked in his cabin, drinking wine straight from the bottle and refusing to speak to anyone. On the morning of the seventh day of the voyage, July 21, 1838, as the ship approached Charleston, Maelzel was found dead in his bed. He left behind only his automata, twelve gold doubloons loaned to him by Alvarez, a few personal papers, a gold medal he had been given many years before by the king of Prussia, and his chessboard. In accordance with naval tradition, a four-pound shot was fastened to Maelzel's feet, and the prince of entertainers was consigned to the deep.

*

The news of Maelzel's death was received with genuine sadness in Philadelphia, where he had been fondly regarded as one of the city's most colourful characters. Maelzel's obituary in the *United States Gazette* lamented that "his ingenuity seemed to breathe life into the work of his hands, but it could not retain the breath in his own nostrils; the kindly smile that he had for children will be no more lighted up on earth; and the furrow of thought that marked his brow as he inspected the movements of the famous Turk, will no more convey intelligence. He has gone, we hope, where the music of his Harmonicons will be exceeded."

Writing in 1859, the Philadelphia chess historian George Allen noted that "to this day, the memory of Maelzel is cherished with a feeling of affectionate respect among those who knew him here, whether merely by attendance upon his exhibitions, or by personal intercourse with himself. His position, in exhibiting for his own emolument the productions of his unsurpassed genius for curious mechanism, was felt to be hardly less dignified than that of the great painter who derives profit from opening a gallery of his own works, or of the great composer who seeks to derive support from his art by presiding at a concert of his own elaborate symphonies."

Since Maelzel had owed him money at the time of his death, Ohl took possession of Maelzel's property and announced that a public auction would be held to dispose of it. The automata and dioramas were to be sold just as Maelzel had left them, packed into their crates. The auction was held in Philadelphia on September 14, 1838, and the Turk was the first item to go under the hammer. It was bought by Ohl himself for the sum of $400. Ohl had no intention of exhibiting the Turk; he bought it as an investment, expecting to be able to sell it for far more to another buyer. But when no offers materialized, he realized that without an expert showman like Maelzel, the Turk was worthless. Eventually, in the spring of 1840, he sold the Turk for $400 to Dr. John Kearsley Mitchell, a prominent local doctor who was a professor at the Jefferson College of Medicine.

As well as having known Maelzel vaguely, Mitchell was the Poe family's doctor. (Edgar Allan Poe and his family had been living in Philadelphia since 1838 and borrowed money from Mitchell on several occasions.) Mitchell had always been fascinated by the Turk and seized the opportunity to discover, at last, how it had actually worked. Did its operation depend on hidden buttons, or magnets? Had his friend Poe guessed correctly, or was de Tournay's article nearer the mark? Mitchell was afraid that if he did not buy the Turk, it might vanish into obscurity. If its secret was not discovered and preserved, he feared, it might be lost forever. To raise the money for its purchase, Mitchell devised a novel scheme. He founded a club whose members would each pay $5 or $10 to join, in return for being let in on the secret. After signing up seventy-five subscribers and collecting $500, Mitchell eagerly took delivery of the five crates containing the dismembered Turk.

It soon became apparent that the automaton was still reluctant to give up its secret. The five crates were found to contain pieces of Maelzel's other automata, including the Carousel, mixed in with the components of the Turk, several pieces of which were also missing and had to be retrieved from other boxes in Ohl's warehouse at Lombard Street Wharf. The Turk had lost its clothes, legs, pipe, and part of its head; a list drawn up by Mitchell of "things to be found and looked for" also included the cabinet doors, castors, and chessmen. This mix-up may have been an intentional stratagem of Maelzel's to prevent anyone from discovering the automaton's secret while it was packed up.

Over the summer of 1840 Mitchell spent his spare time studying the various accounts of the Turk's presentation and theories about its operation. In conjunction with several of his friends, and after "many amusing failures" as his son, Silas, later recalled, he discovered the Turk's secret and restored the automaton to full working order. Finally, during the first week of September, in the unusual setting of Mitchell's own office, the Turk gave its first performance in over two years. Mitchell did his best to ensure that this demonstration resembled the Turk's usual performance as closely as possible, though he himself lacked Maelzel's presence and charisma. And, of course, the demonstration differed from all of the Turk's previous appearances in another crucial respect: once it was over, the secret was carefully explained to those present, who were allowed to examine the automaton at their leisure.

Further performances followed over the next few weeks before the families and friends of the club's members. But once they had been told the secret, their interest soon waned, and Mitchell wanted his office back. The club decided to donate the automaton to the Chinese Museum, a collection of curiosities belonging to a Philadelphian named Willson Peale. Once installed in its new home, the Turk was asked to give a few public performances, and these took place in late November and early December. On one occasion Mitchell, who was the visiting doctor at a local girls' school, invited the pupils to see the automaton play. Although the schoolgirls crowded right around its cabinet and peered into its various doors as they were opened and shut, they could not discover its secret, much to the doctor's amusement.

Without Maelzel's showmanship, and in much reduced circumstances, the Turk had become a pitiful shadow of its former self. Before long the trickle of visitors dried up altogether, and the automaton was moved to a dark corner near one of the museum's back staircases, where it was gradually forgotten.

Fourteen years later, on the evening of July 5, 1854, a fire broke out in the National Theater, on the same street as the Chinese Museum. The flames spread rapidly, engulfing a dozen shops and houses, and by half past ten had reached the museum. There would have been time to rescue the Turk, had anyone thought of it; but by the time Silas Weir Mitchell arrived on the scene to save the automaton, just as his father had a few years earlier, it was already too late. "Struggling through the dense crowd, we entered the lower hall, and passing to the far end, reached the foot of a small back staircase," he later recalled. "The landing above us was concealed by a curtain of thick smoke, now and then alive, as it were, with thick tongues of writhing flame. To ascend was impossible. Already the fire was about him." Standing helpless at the bottom of the stairs, the younger Mitchell paused for a few moments before turning to leave the building. And amid the fire's crackling wood and shattering glass, he fancied that he heard the Turk's last words – *"Échec! Échec!"* – as it was finally engulfed by the flames.

CHAPTER ELEVEN

The Secrets of the Turk

EXPOSED KING: A king with few of its own pawns or pieces close at hand and thus vulnerable to attack.

When you have eliminated the impossible, whatever remains, however improbable, must be the truth.

– Sherlock Holmes, in Sir Arthur Conan Doyle, *The Sign of Four*

Of the many attempts to guess the Turk's secret that were published during its eighty-five-year lifetime, none was entirely accurate. So how did it work? Racknitz guessed a part of the answer, as did Willis, though both of their explanations contained serious flaws. And the few articles that correctly revealed aspects of the Turk's mechanism were not recognized as such at the time, coming as they did among a torrent of speculation that ranged from the plausible to the ridiculous.

It was only in 1857 that an authoritative account appeared, written by Silas Weir Mitchell, whose father had been the Turk's last owner. Mitchell's description of the Turk's secrets took the form of a series of articles titled "Last of a Veteran Chess Player", which were published in *Chess Monthly,* a New York magazine. The account was based on Mitchell's own recollections and some notes made by his father. It repeats a number of myths about the Turk (such as its having played against George III and Louis XV) and contains several errors relating to the manner of the Turk's presentation. But Mitchell's articles, together with other documents dating from the Turk's last days in Philadelphia, make possible a full explanation of the automaton's secret.

As had been widely suspected, the Turk was indeed controlled by an operator concealed inside the cabinet, who remained there throughout the performance. There was no need for wires or pieces of catgut, nor for trapdoors beneath or behind the automaton. Nor was the automaton's strategy guided in any way by the artful use of exterior magnets. Indeed, the exhibitor outside the cabinet, whether it was Kempelen, Anthon, or Maelzel, had no direct control over the automaton's actions. Kempelen's casket, into which he looked during games, and Maelzel's habit of moving his fingers in his pockets or drumming his fingers on the Turk's cabinet were merely diversionary tactics, as many observers correctly surmised.

Willis's explanation of how the operator remained concealed before the game began, which was subsequently borrowed by Poe, was almost entirely correct. The clockwork machinery visible on the Turk's left-hand side (as seen by the audience) did not, as Willis suspected, extend all the way to the back of the cabinet, but only one third of the way. The top of the cabinet lifted off and enabled the operator to climb in and sit on a sliding seat behind this machinery, with his legs extending along the back of the cabinet, behind the drawer, which (as Racknitz had guessed) did not extend all the way to the back of the cabinet either. The back of the drawer was in fact collapsible, so that when the drawer was pulled out, it appeared to have the same depth as the cabinet. Racknitz wrongly concluded that the operator had to fit solely into the space behind the back of the drawer; with the additional space behind the machinery, there was even enough room for the operator to sit up. So there was no need for the operator to be a child, a dwarf, or an amputee; the cabinet was capable of concealing a full-size adult.

At the start of the performance, the operator moved forward on the sliding seat (which was mounted on greased iron rails running along the back of the cabinet), raised his knees, and leaned forward, assuming a somewhat uncomfortable position. The movement of the seat caused a small amount of dummy machinery to move into position behind the clockwork machinery visible through the cabinet's leftmost door (as seen by the audience). It was then possible for the exhibitor to open the small door at the back of the cabinet and hold a lighted candle up to it, whose flickering could just be seen through dense machinery that now seemed to extend all the way to the back of the cabinet.

Once the exhibitor had removed the candle and shut the rear door, the operator straightened his legs and slid backward on the moving seat. This caused the dummy machinery to fold up and also closed a small window behind the frontmost machinery through which the light of the candle had shone. This ensured that there was no danger that the operator, now sitting up behind the clockwork machinery that was still visible to the audience, would be seen.

The operator now had to prepare the cabinet's main compartment for inspection by the audience. As Willis had suspected, the main compartment, which was lined with green baize, was not as simple as it seemed. After sitting back, the operator covered his legs by folding over a lid that formed part of the main compartment's floor; then he concealed his body by closing a door that formed part of the main compartment's side. This ensured that the main compartment resembled an almost empty box. By pulling on a string, and hooking its end over a button, the operator then raised into place the small amount of machinery, including wheels, cylinders, and brass quadrants, that was situated in the main compartment. Again, this machinery was a decoy and had no useful function.

All of this took a few seconds, however, so the exhibitor did not immediately open the cabinet's front doors. Instead, he opened the drawer and slowly and deliberately removed the chessmen. Having done so, the exhibitor could then open the doors of the main compartment and reveal its almost empty interior. Next, he would open the door at the back of the main compartment and introduce a candle, so that the spectators could inspect the main compartment. The exhibitor then spun the whole contraption around and opened the doors in the back of the Turkish figure to reveal more decoy machinery.

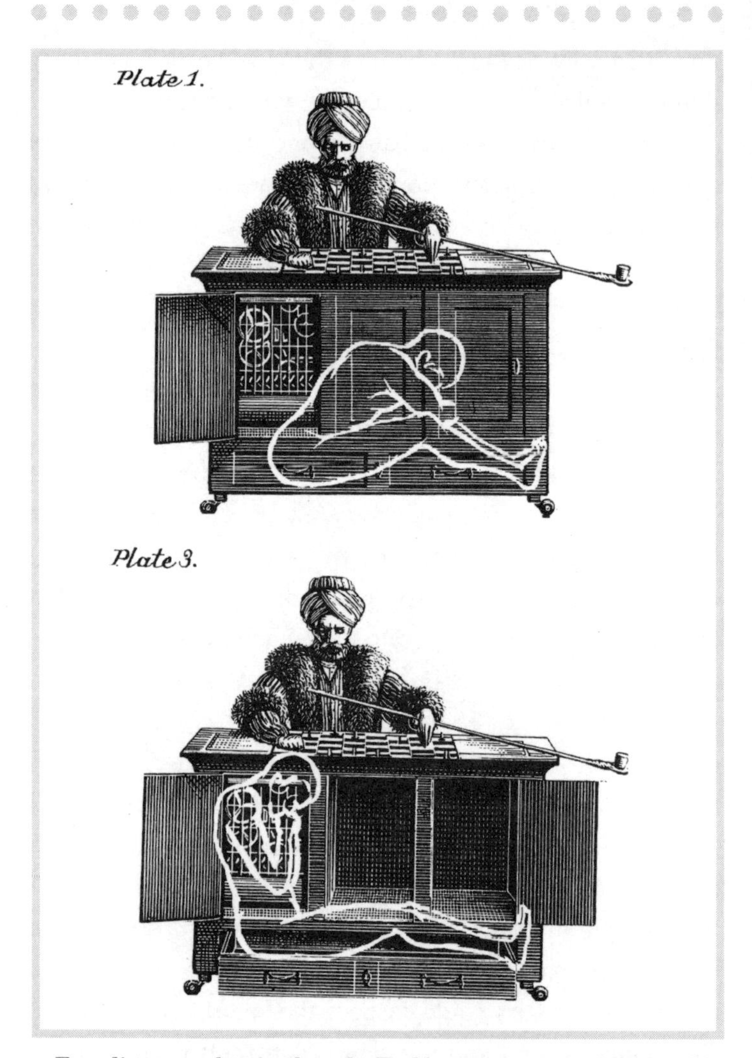

Four diagrams showing how the Turk's operator concealed himself by moving back and forth on a sliding seat and opening and closing various folding partitions.

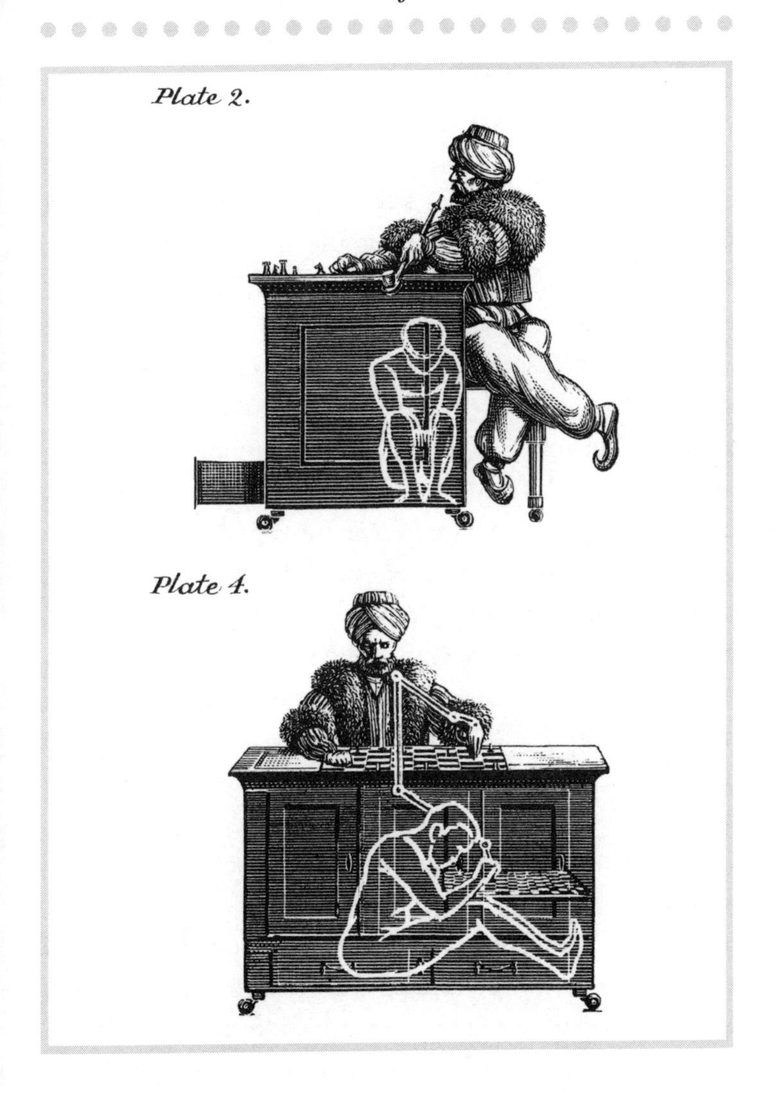
Plate 2.
Plate 4.

Once the audience had been convinced that there was no possibility that anyone could be concealed within the cabinet, the exhibitor could close all the doors, position the cushion under the Turk's left arm, remove its pipe, and wind the automaton up. During this time the operator rearranged the Turk's interior once again. The door to the main compartment was opened and its decoy machinery folded away. Next, the operator opened a small door to his left, revealing a tiny compartment in the Turk's body containing a burning candle that gently illuminated the cabinet's interior. The smoke from this candle was carried up a chimneylike pipe to an aperture in the top of the Turk's turban. The purpose of the two candelabra that were placed on the cabinet during the Turk's performances was to disguise the smell of burning wax and the smoke from this interior candle. Small airholes around the top of the cabinet provided fresh air for the operator. Even so, the inside of the Turk was dark, hot, and smoke-filled, which was why its performances had to be limited to an hour or so; no operator could stand to remain inside the automaton for any longer.

The operator fixed a chessboard in front of him, each of whose squares was perforated with two round holes. One hole was used to secure the operator's chessmen, whose undersides were equipped with suitable pegs. The second hole was used, in conjunction with a set of levers called a pantograph, to position the Turk's arm with great precision.

Willis was wrong to assume that the Turk contained no genuine machinery at all, and that the operator simply guided the Turk's arm by putting his own arm inside it. Instead, as Racknitz had correctly guessed, the operator controlled the Turk's arm using a highly sophisticated mechanism.

The operator grasped a metal pointer, which could be moved to point at any square on the internal chessboard. The pointer was connected to the Turk's arm via a system of levers, which were arranged so that whenever the pointer was positioned over a particular square, the Turk's hand would be positioned over the corresponding square of the external chessboard. The pointer could also be moved up and down, which caused the Turk to raise and lower its hand. And by twisting the end of the pointer, the operator could open and close the fingers of the Turk's gloved left hand. Each time he made a move, the operator moved the appropriate piece on his internal chessboard and then directed the Turk's arm to move the corresponding piece on the external board. If the chessman slipped from the Turk's fingers, the operator would complete the move to indicate to the exhibitor the intended destination square of the chessman. The exhibitor would then pick up the fallen piece and place it on the appropriate square.

But how could the operator tell how his opponent had moved? Willis wrongly asserted that the operator was hidden inside the Turkish figure and could see the external chessboard by looking through a gauze-covered aperture in its chest; no internal chessboard would then have been needed. The reality was more complex. As Racknitz and Böckmann had suggested, the chessmen on the external board contained small but powerful magnets; and just under each square of the board was a small metal disc, suspended on a delicate coiled wire. When a chessman was put down on a particular square, the disc underneath was attracted by the magnet hidden inside the chessman and moved upward to make contact with the underside of the board. When the chessman was picked up, the disc fell away and wobbled for a few seconds on its coiled wire. By watching the underside of the board, first for a descending and wobbling disc, and then for a disc suddenly moving upward, the operator could work out which chessman had been moved on the external board. He could then make the corresponding move on his internal board and start considering his response.

So the Turk's mechanism did depend on magnetism after all, at least to some extent. Placing a large magnet on the top of the cabinet, however, evidently had no effect on the metal discs under the chessboard, since Kempelen and Maelzel allowed it on several occasions. And wrapping a shawl around the Turkish figure's torso, as Napoleon was alleged to have done, would have had no effect on the operator's ability to follow the game.

As well as using the rising and falling discs to follow his opponent's moves, and using the pantograph to make his own, the operator had a few other gadgets at his disposal. There was a deliberately noisy piece of clockwork which did nothing, but which could be started and stopped at will by the operator. Having decided on his move, the operator would start this clockwork running, make his move using the pantograph, and then stop the clockwork. This heightened the impression that the movements of the Turk's arm were governed by the intricate clockwork machinery that had been revealed at the beginning of the performance. In addition, the operator pulled strings and levers to cause the Turk to shake its head, roll its eyes, rap its right hand on the table, and activate its voice box to say "*Échec*". There was a noisy, rattling spring that could be activated if the operator needed to conceal the sound of a cough or sneeze. The operator also had a perforated paper diagram indicating the sequence of moves required to perform the Knight's Tour, which was placed over the Turk's internal chessboard when needed.

Finally, there was a simple device to enable the operator and exhibitor to communicate in the event of an emergency. It consisted of two small circular discs of brass, mounted opposite each other on the inside and outside of the cabinet's rear wall, and numbered with the digits 0 to 9. A thin rod passed through the middle of these two discs, on either end of which was a pointer, which could be rotated either by the operator or by the exhibitor and used to indicate a particular number, whose meaning would have been agreed on in advance. The exhibitor could thus instruct the operator to let his opponent win, or hurry up, or make less noise, and the operator could tell the exhibitor that he was ready to start the game, or wished to abandon the game, or that his candle needed relighting (which would be done under the pretext that the exhibitor needed to adjust the Turk's machinery).

The folding compartments, the pantograph, the magnetized chessmen, the rising and falling discs – all these elements of the Turk's operation had been suggested by various observers over the years, but no external observer ever quite put all the pieces together correctly. What is perhaps even more surprising is that so little was ever revealed by the select band of men and women who operated the Turk during its extraordinary career.

*

The identity of the Turk's original operator, who was hiding inside the automaton on the occasion of its debut in Vienna in the spring of 1770, is unknown. There have been suggestions that it was one of Kempelen's children, but at the time he had only a daughter, Theresa, who was about two years old. It is certainly possible that one of Kempelen's children operated the Turk on its European tour, in the 1780s; by this time Theresa was in her mid-teens, and Kempelen's son, Carl, who was born in 1771, was twelve or thirteen. But it is most likely that there were several operators during Kempelen's lifetime. Some evidence for this can be found in Kempelen's letter to Sir Robert Murray Keith, the Scottish nobleman who saw the Turk play at Kempelen's home in August 1774. Kempelen claimed that the automaton had been damaged on the way to a recent performance and that he would need a week or two to fix it. He also admitted that it had played very badly, "as you have probably heard". Presumably the Turk's operator at the time was not a very strong player, and Kempelen wanted to buy some time in which to find and train someone better. He was clearly able to engage a strong player for the Turk's European tour, during which the automaton took on some of the best players in Europe and lost only to the very best. But exactly who operated the Turk during Kempelen's lifetime remains a mystery.

The first of the Turk's operators whose name is known is Johann Allgaier, a Viennese chess master who operated the automaton on Maelzel's behalf until he sold it to Eugène de Beauharnais. Allgaier would therefore have been the operator during the Turk's match with Napoleon. It is said that Prince Eugène employed a young girl to operate the Turk while it was in his possession; but as with the theory that Kempelen's children operated the automaton, this may simply be a reflection of the once-prevalent belief that only a child or dwarf could possibly have fitted inside it.

More is known about the Turk's operators from 1818 on, when Maelzel regained possession of the automaton and took it to Paris. Instead of challenging the players at the Café de la Régence to play against the Turk, Maelzel hired them as its operators. First to climb inside the automaton was Boncourt, a strong, reliable French player who was also very tall and therefore unlikely to be suspected as the Turk's motive force. But Boncourt was a slow player, and on several occasions he almost let the cat out of the bag by sneezing during a game. This prompted Maelzel to install the noisy spring, to cover up any future coughs and sneezes. While in Paris, the automaton was also operated by two other players: Aaron Alexandre, who was well known as the author of a chess encyclopedia, and a man named Weyle about whom little is known.

When the Turk moved on to London, Maelzel hired William Lewis as its operator. This young man was a student of Jacob Sarratt, the best player in England, who had turned down the opportunity to operate the Turk on the grounds of his advanced age. On one occasion when Lewis was inside the automaton, he realized that his opponent was an unusually gifted player and began to suspect that it was Peter Unger Williams, another of Sarratt's students. After an epic game, lasting an hour and a half, Lewis eventually triumphed and subsequently discovered that his opponent was indeed Williams. Lewis convinced Maelzel to let Williams in on the secret, and soon he too was operating the automaton.

In the summer of 1819, when Maelzel proposed a tour of

the north of England, he had to find a new operator since neither Lewis nor Williams was prepared to leave London. Eventually he settled on a replacement: a Frenchman named Jacques-François Mouret, who was Philidor's grand-nephew. Mouret was used to playing quickly, as was the custom in Paris, which gave him an advantage over English players, who preferred to take their time. By rapping the Turk's right hand on the cabinet and rolling its eyes, he could put pressure on his opponents to hurry up. Mouret seems to have remained the Turk's operator until Maelzel departed for America at the end of 1825.

In his haste to leave Europe, Maelzel set out for America without having engaged a new operator, and this explains why the Turk initially played only endgames. During the voyage across the Atlantic, Maelzel was forced to teach a young Frenchwoman, who was a complete novice to the game of chess, both the rules of the game and how to operate the automaton. In particular, she concentrated on learning the endgames in Maelzel's little green book, several of which had been carefully chosen to guarantee victory for whichever player moved first. Since the Turk always moved first when playing endgames, victory was assured much of the time; the rest of the time, the young woman's familiarity with the endgames would have given her an advantage. When the woman was suspected of being the automaton's operator during its first run in New York, Maelzel arranged for a young man to take her place so that she could be seen sitting in the audience and playing against the automaton, and thus dispel suspicion. The young man in question was the son of Maelzel's ally William Coleman, editor of the *Evening Post*.

The Turk's sudden ability to play full games while in Boston in October 1826 was due to the arrival from Europe of William Schlumberger, a regular at the Café de la Régence, whom Maelzel had employed as the automaton's new full-time director. Schlumberger quickly became Maelzel's most trusted associate and acted as his secretary and right-hand man; he was widely suspected of being the Turk's operator. It was Schlumberger who was seen getting out of the Turk's cabinet by the two young boys in Baltimore, whose story had of course been true all along, and Poe rightly pointed out that when Schlumberger was unwell, exhibitions of the Turk were suspended. Schlumberger remained the Turk's director for the rest of his life; his untimely death was a double blow to Maelzel, for whom Schlumberger came to be seen as the son he never had, as well as the life force of his beloved automaton.

After Maelzel's own death the Turk was operated by several people, but most often by Lloyd Smith, a young Philadelphian whose father had been one of the members of the club formed by John Mitchell to buy the automaton. Smith accompanied his father to a demonstration of the automaton and was invited to operate it by Mitchell, who was, Smith later recalled, "having some trouble with the man inside". He immediately took over as the Turk's operator and gave several performances at Mitchell's offices. Since he was also inside the Turk for its handful of performances at the Chinese Museum, Smith was very probably the last person to operate the automaton.

In 1858 Smith wrote to the chess historian George Allen and gave a brief account of his experiences operating the Turk. His letter, now in the archives of the Library Company of Philadelphia, represents the only firsthand explanation of the Turk's secret and even includes a few scrawled sketches showing the various positions assumed by the operator. Smith told the story of one of the Turk's last games, in which he "found his match in a Mr Chase, and seeing I must be beaten I sprung the machinery . . . and communicating with the exhibitor . . . he told the audience that the machinery had got out of order and the exhibition must unavoidably be postponed."

Several other amusing tales, of varying degrees of credibility, are told about the Turk's operators. On one occasion when Maelzel was displaying the Turk in a small town in Germany, presumably with Mouret as its operator, a local conjuror felt that he was being upstaged and decided to sabotage the automaton's performance. Convinced that there had to be a man inside the cabinet, the conjuror and an accomplice shouted "Fire! Fire!" while the Turk was playing, prompting the spectators to flee in panic. At the same time, the automaton started to shake violently, and Maelzel only just managed to wheel it out of sight before its operator revealed himself.

Another dubious story tells of the Turk's visit to Amsterdam, where Maelzel was supposedly offered a vast sum to stage a demonstration before the king of Holland. Mouret, who had not been paid by Maelzel for some time, announced on hearing the news that he was suddenly too ill to play. Only by handing over the overdue wages could Maelzel get Mouret to assume his position inside the automaton. The king nominated his minister of war to play on his behalf but advised him from the sidelines; as is usually the case in such stories, the two men were duly defeated.

Of the Turk's many operators, Mouret seems to have been the only one who ever betrayed its secret. The article that appeared in *Le Palamède,* a French chess journal, in 1836 (which was reprinted in the *National Gazette* in America in 1837), contained information that with hindsight could only have been known to someone who had actually operated the Turk. The *Palamède* article was itself based on an earlier article that appeared in the *Magazine Pittoresque* in Paris in 1834. On the surface, this article was simply a rehash of Racknitz's theory, since it referred to magnetized chessmen, two chessboards, and so forth. But it also referred to the operator sitting "on a shelf with castors", a clear reference to the sliding seat that actually existed inside the Turk. The most likely source of this information was Mouret, who was living in Paris at the time the article appeared. Evidently he had fallen on hard times and had sold his story; the chess historian George Walker, writing in 1839, recorded that "he burnt out his brain with brandy, and died recently in Paris, reduced to the extremest stage of misery and degradation."

As it turned out, Mouret's disclosures had little impact. Maelzel and the automaton had long since gone to America, where the article did not appear until 1837, by which time the Turk's fortunes were already in decline. Although Maelzel did not respond when the *National Gazette* article appeared, he cut out the article and kept it, and it was found among his papers after his death.

*

Despite Silas Weir Mitchell's detailed account of the Turk's secret and Lloyd Smith's letter corroborating it, the incorrect explanation advanced by Willis and Poe has cast a long shadow over subsequent descriptions of the automaton. The theory that the Turk's operator ascended into the figure's torso and watched the game through a gauze screen seemed the most plausible explanation to many later writers, who considered the magnetized chessmen and elaborate pantograph to be overly complicated. Could Kempelen really have built such a complex mechanism and got it to work so reliably? A number of sceptics expressed their doubts.

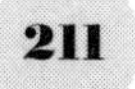

In his 1949 book *A Short History of Chess,* for example, Henry A. Davidson dismissed the idea of magnetized chessmen and so on as unlikely, "since knowing that a square was occupied could not tell the unseeing operator which piece rested on that square. Poe's explanation is more credible. He thinks that during the game, the player sat inside the figure and looked out at the board through the gauze bosom of the Turk." This argument is flawed, however. It is true that the magnetized chessmen and metal discs did not directly indicate the identity of the chessman that had been moved, just its initial and final positions; so if the chessmen on the external board had been positioned at random, there would have been no way for the operator to determine what was going on. But whenever the Turk played, the exact starting configuration of the chessmen was known to the operator, whether for a full game or an endgame, and could be duplicated on the internal chessboard. So each time a chessman was moved on the external board, merely knowing its initial and final position gave the operator enough information to determine its identity, by consulting his own board.

In his comprehensive and detailed book *Chess: Man vs Machine,* published in 1980, Bradley Ewart declared the Willis/Poe explanation to be "the most reasonable". Ewart pointed out several errors in Mitchell's description of the order in which the Turk's doors were opened and closed. He also noted that, according to Poe, the Turk sometimes seemed to anticipate its opponent's moves after they had been made on the second chessboard, implying that the hidden operator could see the second chessboard. Ewart even quoted Thicknesse, who had originally suggested the idea of a hidden operator inside the Turk's torso, and who claimed to have seen the Turk's clothing move when the figure should have been quite motionless. Ultimately he concluded that "this one secret of the Turk remains a mystery."

Alex G. Bell's book, *The Machine Plays Chess,* published in 1978, offered yet another incorrect explanation. Ignoring the mountain of evidence to the contrary, Bell wrote that he thought it "unlikely that it was necessary or desirable for the chess master to actually be inside the machine itself. It is far more likely that the operator was a trained boy (or very small adult) who followed the directions of the chess-player who was hidden elsewhere on the stage or in the theatre – the Turk was a 'mind-reading' act. Nowadays the techniques of passing information non-verbally are better known. The simplest code for the Turk's operator would be a set of signals for start, left, right, up, down and stop. These signals can be made in a variety of ways (hand in pocket, stance, head movements, etc.) . . . it is certainly difficult to believe that any human who must perform efficiently (and therefore as comfortably as possible) would subject himself to the confines of the Turk's chest, particularly if a boy could hide himself more easily within the machine. Unfortunately we shall never know."

One way to show that the magnetized chessmen and

pantograph system could indeed have worked as described would be to reconstruct the Turk. And in 1971 that is what John Gaughan, a prop builder for professional magicians and a collector of automata, began to do.

The workshop in Los Angeles where Gaughan and his assistants construct their elaborate stage machinery gives an insight into the circumstances in which Kempelen and Maelzel built their automata. There are wood- and metal-working tools, and workbenches strewn with components and partly assembled machinery. Planks of wood and sheets of metal are stacked up against the walls, which are adorned with huge posters advertising magicians of the past. A talkative parrot preens itself on a wooden perch. An automaton clarinet player, currently being restored, stands immobile in one corner; its innards, including a brass-studded wooden drum and a set of small bellows, closely resemble those of Maelzel's automaton trumpeter. A passage leads to Gaughan's dark, book-lined study. On the wall ticks a wooden clock, handmade by Gaughan himself. Farther down the passage in a storeroom sit more automata, including a flowering orange tree, a writing figure, a trapeze artist, and a cardplayer. And in one corner, staring straight ahead as though deep in thought, sits a familiar figure: a perfectly faithful reconstruction of the Turk.

Gaughan first became interested in the Turk as a young boy, when he came across a reference to it in a book about the history of magic. In 1971, when he set up his prop-

John Gaughan's reconstruction of the Turk.

making business, he decided to reconstruct the automaton. He started off with a wooden box, in which one of his assistants assumed the various positions of the Turk's operator; the construction of the automaton then proceeded from the inside out. Over the next eighteen years Gaughan started again several times, refining the design each time.

During the construction, Gaughan was surprised to find

that getting the pantograph to work was fussy, but was not as difficult as he had expected. He also reconstructed the magnetized chessmen and metal discs, and found that the discs can be set up so that they wobble for as long as thirty seconds each time a chessman is moved. This means the operator need not worry about missing his opponent's move and does not have to stare at the underside of the board continuously. Gaughan figured out the arrangement of the folding partitions inside the cabinet and gave his Turk the ability to roll its eyes, turn its head, and rap on the table, just as the original automaton did. Eventually, in November 1989, the reconstructed Turk was ready to give its first performance, at a conference in Los Angeles on the history of magic.

The new Turk was presented in the same manner as the original, with one or two small changes. Like Maelzel, Gaughan did away with Kempelen's mysterious casket. And rather than illuminating the Turk's interior with a candle, the operator used an electric light, so that Gaughan could dispense with the candelabra that were needed to disguise the smoke emanating from the Turk's interior. Finally, Gaughan added two touches of his own: He suspended a large mirror at an angle behind and above the Turk, so that the audience could see the chessboard and follow the game more easily. And on the stage Gaughan set up a personal computer, running a chess program, to be the Turk's opponent. The Turk challenged the computer to an endgame, carefully chosen so that whoever moved first would be sure to win; since the Turk had the first move, it quickly defeated the computer. "It was wonderful," Gaughan recalls.

As an expert on the history of magic, and having built his own Turk, Gaughan has a unique perspective on why it was so successful. Notwithstanding the complexity of the pantograph, Gaughan's view is that the Turk is primarily an example of a magician's rather than an engineer's ingenuity. Knowing how the Turk's insides work mechanically is only part of the explanation, Gaughan insists; to fully explain the effect it had on people, it is necessary to appreciate the magician's stagecraft involved in its presentation.

For starters, he points out, Maelzel used to display the Turk at the end of a show that consisted of several genuine automata, to get the audience into the right frame of mind, so that they assumed that there was no limit to what mechanical contraptions could do. Next, the initial display of the Turk's clockwork machinery, which appears to fill one-third of the cabinet, is intended to fool anyone who suspects there to be an operator inside the cabinet into thinking that the operator must be hidden in the other two-thirds. This perception is, however, totally undermined when the doors to the main compartment are opened, revealing it to be almost empty. Misdirection, the technique of subtly suggesting an explanation to the spectators and then undermining it, is common to many tricks that involve hiding people, animals, or things inside apparently empty containers. (Such tricks are known as "cabinet illusions", of which the Turk is regarded by historians of magic as the first.) Gaughan also believes that the practice of leaving the cabinet's doors open, so that they flapped about as the Turk was rotated on its castors, is a key part of the illusion. Letting the doors move around gives the exhibitor an air of studied casualness, implying that he has nothing to hide; closing all the doors firmly, on the other hand, would give the impression that the exhibitor wishes to control what the audience can or cannot see. Furthermore, following the engravings of the original Turk in Windisch's pamphlet, Gaughan found that one of the rear doors is hinged in such a way that, left to itself, it remains slightly ajar. This provides the operator with much-needed light and fresh air during the performance.

Another important aspect of the illusion is the loud noise made when winding up the automaton. Gaughan has concluded that the automaton's main key was probably a crank handle that made a noisy ratcheting sound as it was turned. This noise, and the physical effort required to turn the handle, would have strengthened the spectators' belief that the clockwork machinery shown to them before the game began was responsible for determining the Turk's moves and guiding its arm.

It is little touches like these, says Gaughan, that explain why the original Turk was so convincing and successful. What is perhaps his greatest discovery is that when watching his reconstructed Turk, the illusion that it is genuinely a pure machine is extraordinarily compelling, even to those who know how it works. Something about the Turk seems to evoke a fundamental human desire to be fooled. "The presenter dangled the carrot, and didn't try to say it worked like this or that," says Gaughan. "That's very good. That's strong magic. That's good theatre."

Through his reconstruction of Kempelen's automaton, Gaughan has shown that Mitchell's explanation of its mechanism is correct, and that the combination of pantograph and magnetized chessmen really works. He has thus extinguished any remaining doubts about the nature of the original Turk's mechanism.

*

In the Turk's early career, the idea that it might have been a pure machine evidently did not seem entirely out of the question, since a number of quite learned observers were prepared to believe it. The late eighteenth century was a time when the possibilities offered by mechanical devices seemed limitless; to some people, a chess-playing machine would have seemed only slightly more miraculous than a steam engine or a power loom.

By the end of the Turk's career, however, the idea that it might have been a genuine automaton seemed far less credible. Complex machines from railway engines to electric telegraphs were becoming a part of everyday life, and the people of the mid–nineteenth century considered themselves better able to evaluate what machines could and could not do. Once the Turk had been unmasked as a pseudoautomaton, rather than a genuine "self-moving machine", the idea of a genuine chess-playing machine suddenly seemed absurd. To drive the point home, in 1879 an automaton maker named Charles Godfrey Gümpel produced a "proof" that a chess-playing machine had been impossible all along and, furthermore, always would be. (Gümpel was the creator of Mephisto, one of several pseudoautomaton chess players that were constructed after the Turk's demise.)

Gümpel estimated the number of possible configurations of chessmen on a chessboard to be 10^{32}, or 100,000 billion billion billion. He then imagined a chess-playing machine that would use punch cards, of the kind used in a Jacquard loom, to encode the correct move to be made in each position. Ignoring the time required to work out what the right move would be, and glossing over the internal design of the machine, Gümpel simply considered how long it would take to drill one hole for each possible configuration. He assumed that a workman could drill a hole every three seconds, and concluded that a man working ten-hour days, 300 days a year, for fifty years, could drill 180 million holes in his working life. The machine would therefore require 500,000 billion billion people to devote their entire working lives to drilling the holes; such a machine was, in other words, entirely impractical.

Richard Proctor, a nineteenth-century English science

writer, noted that even if the punch cards could be drilled, they would cover an area about 1 million times larger than the floorspace of London's Crystal Palace, the site of the Great Exhibition of 1851. And since Kempelen had built his automaton in just a few months, Proctor concluded, he could not possibly have accomplished "all that was requisite to make a true automaton player".

This argument sounded foolproof. It seemed obvious that a chess-playing machine would never be built. But the events of the twentieth century were to prove this assertion to be completely erroneous.

CHAPTER TWELVE

The Turk Versus Deep Blue

DISCOVERED CHECK: Check given by a queen, rook, or bishop when another chessman is moved out of the way. If the moved chessman also gives check, the result is a double check.

Any sufficiently advanced technology is indistinguishable from magic. – Arthur C. Clarke

The confident nineteenth-century predictions that a chess-playing machine could never be built were, of course, quickly disproved following the invention of the digital computer. Computers are unquestionably the modern descendants of automata: they are "self-moving machines" in the sense that they blindly follow a preordained series of instructions, but rather than moving physical parts, computers move information. Furthermore, just like automata before them, computers operate at the intersection between science, commerce, and entertainment. And they have given rise to an industrial revolution of their own, by extending human mental (as opposed to physical) capacity.

Curiously, the first genuine game of chess between man and machine did not involve a computer, at least not in the modern sense. The match took place in 1952 in Manchester, England, in a first-floor office at the Royal Society Computing Laboratory. The sparsely furnished office contained a few chairs and a small table, covered with untidy piles of paper. The human player was Alick Glennie, a twenty-six-year-old computer researcher, who was learning computer theory in order to apply it to the British government's secret atomic weapons project. Playing the part of the computer was the British mathematician and computer scientist Alan Turing.

In 1937 Turing had published a landmark article in computer science, "On Computable Numbers", in which he demonstrated that the solutions to some kinds of mathematical problems cannot be calculated by machinery, no matter how complex or powerful that machinery is. Turing was interested in the theoretical limits of computing machinery, though electronic computers did not exist at the time. (The word *computer* referred to a mathematician skilled at performing fast, accurate calculations.) Turing was familiar with Charles Babbage's earlier work, and, like Babbage, he was particularly interested in programming machines to play chess. He was not alone among computer scientists in regarding chess both as a first step toward developing an intelligent machine and as a means of comparing human and machine intelligence.

In America, the computer scientists John von Neumann and Oskar Morgenstern were also thinking about the possibility of programming computers to play chess, and in 1950 another American computer pioneer, Claude Shannon, published the article "A Chess-Playing Machine", in which he advocated the construction of such a device. "The investigation of the chess-playing problem is intended to develop techniques that can be used for more practical applications," he wrote. "The chess machine is an ideal one to start with for several reasons. The problem is sharply defined, both in the allowed operations and the ultimate goal. It is neither so simple as to be trivial or too difficult for satisfactory solution. And such a machine could be pitted against a human opponent, giving a clear measure of the machine's ability in this kind of reasoning."

During World War II, Turing's interest in computer chess had been common knowledge among his code-breaking colleagues at Bletchley Park, the English country house where the Nazis' codes were cracked through a combination of human ingenuity and specialized but primitive computer technology. One of Turing's associates recalled that "some of our discussions were concerned with the possibilities of machine intelligence, and especially with automatic chess-playing. We agreed that the most interesting aspect of this topic would be the extent to which the machine might be able to simulate human thought processes." In a newspaper report from 1946 in which he discussed the prospects for general-purpose computers, Turing raised the possibility that such machines might eventually be able to play chess.

Indeed, even though he did not have a computer to run it on, Turing had started writing a simple chess-playing program in the 1940s. Like all computer programs, it consisted of a rigid set of rules that were to be followed in a particular order, and could determine what move to make in a particular chess position. These rules were written on half a dozen sheets of paper. In 1952, once the program was finished, Turing was curious to see how it would perform. One day, over lunch, he asked his colleague Glennie if he would like to play a game of chess against this "paper machine". Turing proposed to play the part of the computer himself and work through the program's steps by hand for each move. Glennie agreed, and that afternoon the two men retired to Turing's office, set up a chessboard, and played what is now regarded as a historic game.

In some respects, Glennie was following in the footsteps of the Turk's many opponents, in the sense that he was playing a game of chess against a person who was pretending to be a chess-playing machine. The difference, of course,

Alan Turing

was that unlike the Turk's operators, Turing had no choice over what move to make. Instead, after each of Glennie's moves, Turing performed the laborious task of following the rules of his chess-playing program to determine his response. The program was extremely simple and played very badly. "During the game Turing was working to his rules and was clearly having difficulty playing to them because they often picked moves which he knew were not the best," Glennie later recalled. "He also made a few mistakes in following his rules which had to be backtracked. He had a tendency to think he knew the move the rules would produce and then have second thoughts. He would then try to find the piece of paper containing that section of the rules, and to do so would start juggling with all his papers. We were playing on a small table which did not help."

Glennie was a weak player, but after three hours and twenty-nine moves he defeated Turing's paper computer. Turing was not surprised; he cheerfully admitted that his program, which he had overoptimistically named Turbochamp, played in a way that was a caricature of his own feeble playing style. Turing subsequently started to write a chess program in his spare time to run on an actual computer (a Manchester machine called MADM) but did not get very far with it. He died after ingesting poison on June 7, 1954; whether his death was accidental or deliberate suicide remains unclear.

Although he never got a chess program running, Turing nonetheless became one of the founders of computer chess, and of the field that subsequently came to be known as "artificial intelligence". He is also remembered today as the inventor of a simple test, known as the Turing test, which can be used to determine whether a computer is intelligent or not. Simply put, a computer passes the test and is deemed an intelligent "thinking machine" if a human conversing with it via typewritten messages cannot tell whether it is another human or a machine. Turing proposed this "imitation game" as a yardstick for machine intelligence in an article published in 1950, where he gave a fictitious example of the sort of question-and-answer session he had in mind. It even includes a chess problem, underlining Turing's belief that intelligence, chess playing, and the ability to hold a conversation are closely linked.

Q: Please write me a sonnet on the subject of the Forth Bridge.

A: Count me out on this one. I never could write poetry.

Q: Add 34957 to 70764.

A: (Pause about 30 seconds and then give as answer) 105621.

Q: Do you play chess?

A: Yes.

Q: I have K at my K1, and no other pieces. You have only K at K6 and R at R1. It is your move. What do you play?

A: (After a pause of 15 seconds) R-R8 mate.

Turing proposed that any computer program capable of holding such a conversation, so that a human observer cannot tell if it is another human or a machine, could be considered to be thinking. "I believe that in about fifty years' time it will be possible to program computers to make them play the imitation game so well that an average interrogator will not have more than 70 per cent chance of making the right identification after five minutes of questioning," he declared. Turing thus proposed two yardsticks for machine intelligence: the ability to play chess, and the ability to hold a conversation.

The first chess game between a human and an electronic computer followed a few years later, in 1958. The program in question was written by Alex Bernstein, an American researcher, and it ran on an IBM 704 computer. It evaluated about 3,000 positions before making each move, a process that took around eight minutes. Even so, it was a weak player. But as computers got faster and computer-chess theory evolved, the ability of chess-playing programs improved steadily. Herbert Simon, a pioneering researcher in artificial intelligence, predicted in the late 1950s that the world chess champion would be a computer within ten years.

*

During the 1960s the outlook for artificial-intelligence research was bright. Its ultimate goal was perhaps most clearly encapsulated by HAL 9000, the intelligent computer that appeared in Stanley Kubrick's 1968 film *2001: A Space Odyssey*. In the film, HAL is the brains of the spacecraft Discovery and converses with the craft's human crew in a gently hypnotic speaking voice. According to the novel based on the film's screenplay, written by Arthur C. Clarke, HAL "could pass the Turing test with ease". But talking computers or robots had been a staple of science fiction for some years, so to underline HAL's intelligence, one scene in *2001* depicts a game of chess between HAL and one of Discovery's astronauts. HAL, of course, wins the game easily. Later in the film, HAL further demonstrates the general nature of his intelligence by learning to read lips. (HAL subsequently malfunctions and tries to kill the Discovery's crew because, it later transpires, he has been forced to lie to them about the true purpose of their mission; he decides that the easiest way to resolve the internal logical conflict caused by having to tell lies is to kill the astronauts.)

Kubrick and Clarke went to great lengths to make the future depicted in *2001* as believable and accurate as possible. When the film was released in 1968, it seemed plausible that space hotels, a permanent moon base, and artificially intelligent computers would exist within thirty years. These intelligent computers would be able to pass the Turing test and would also be able to apply their intelligence to a range of other analytical problems, such as playing chess. The ability to converse and play chess were, in other words, depicted as fundamental capabilities of intelligent machines.

But things did not turn out the way *2001* predicted. The space hotels and moon base have yet to materialize, and artificial-intelligence research has also failed to live up to the expectations of the 1960s. The problem, in essence, is that although progress has been made in specific areas of problem solving – computers can now, for example, play chess, fly aircraft, recognize faces, and transcribe continuous speech – there has been little progress in integrating these advances into a generalized theory of machine intelligence. Shannon's optimism that getting computers to solve particular specialized problems would result in techniques that could be more generally applied has proved unfounded. Instead of multiskilled computers like HAL, the result has been systems that are able to mimic some degree of human intelligence when performing a particular task, but which are otherwise clueless. (In the words of computer scientist Anatol Holt, "A brilliant chess move while the room is filling with smoke because the house is burning down does not show intelligence.")

Chess is unquestionably the field in which specialized artificial-intelligence research has made the most progress, but Simon's prediction of a computer world champion by 1970 was too optimistic. The first really formidable computer-chess program was Chess 4.0, written by David Slate and Larry Atkin in 1973. Its ability improved as available computer hardware got steadily faster, and it reached what chess players call "expert" level in 1979. In 1983, researchers at AT&T Bell Labs built a computer called Belle, which relied on special chess-processing chips to analyse positions even more quickly, and which reached "master" level. This approach eventually led to the most powerful chess-playing machine ever built, a machine capable of taking on the human world champion: Deep Blue.

*

In October 1989, as John Gaughan was putting the finishing touches on his reconstructed Turk in Los Angeles, an entirely different chess-playing machine was getting ready for a match of its own. Deep Thought, the predecessor of Deep Blue, had been built the previous year by a team of researchers at Carnegie Mellon University, including Murray Campbell and Feng-Hsiung Hsu. Thanks to its special chess-processing chips, with which it was able to analyse 700,000 chess positions every second, Deep Thought had quickly established itself as the world champion of computer chess. Now it was to meet Garry Kasparov, the twenty-six-year-old Soviet world chess champion, for the very first match between the top human and computer players.

The match, which consisted of two ninety-minute "sudden death" games, took place on October 22 at the New York Academy of Fine Arts. Kasparov, who had repeatedly stated his conviction that a computer would never defeat the world chess champion, felt that he had more than merely his own reputation to defend by playing Deep

Thought. "Of course, I have to challenge it just to protect the human race," he told reporters. As it turned out, he won the first game without much difficulty and defeated the computer even more convincingly in the second. The match received widespread press coverage, and Kasparov became a household name – not least because he was seen as having defended humankind's honour. By emphasizing the man-versus-machine nature of the match, in which he represented all of humanity, Kasparov even managed to transcend the Cold War rivalry between America and the Soviet Union.

The Deep Thought team spent over six years preparing for a rematch. Campbell and Hsu moved to IBM's Thomas J. Watson Research Center in upstate New York, where they redesigned the machine to make it far more powerful. Deep Thought had been a single computer with a handful of additional chess chips; its successor, called Deep Blue, consisted of thirty-two IBM SP-2 supercomputers connected to form a single machine. Each SP-2 contained up to 8 improved chess chips, so that there were 220 chess chips in all. This enabled Deep Blue to evaluate 100 million positions per second.

The first full match with Kasparov, who was still the world champion, took place in February 1996 in Philadelphia. It was to be a six-game match, played under tournament conditions, with a prize fund of $500,000. Astonishingly, Kasparov stumbled in the first game and went on to a historic defeat: it was the first time a computer had defeated the world champion under match conditions. But he came back in the remaining five games, winning three and drawing two, and becoming increasingly dominant as the match progressed. Once again, man had defeated machine. IBM, which had put up the prize money as well as funding the development of Deep Blue, was bound to win either way; the company estimated that media coverage of the game was equivalent to $250 million in "free favourable advertising" for its products and immediately proposed a rematch, to which Kasparov agreed without hesitation.

The rematch took place in New York in May 1997, on the thirty-fifth floor of the Equitable Center in downtown Manhattan. In the building's auditorium, spectators (who paid $25 for tickets to each game) watched the action over closed-circuit television; five of the six games sold out. The prize fund was increased to $1.1 million, of which $700,000 would go to the winner. In the months since the 1996 match, Kasparov's performances against other human players had been the best of his career, and he was widely regarded as the favorite. Deep Blue Junior, a smaller version of Deep Blue, had spent the intervening months touring America playing exhibition games in front of rapt audiences. Meanwhile, Deep Blue itself had been extensively upgraded. Its thirty-two SP-2 supercomputers now contained a total of 512 chess chips, making Deeper Blue, as the machine was sometimes called, twice as fast as it had been the previous year. It was now capable of analysing 200 million positions per second.

Once again, the match received enormous media coverage, much of which emphasized the notion that Kasparov was defending humanity's honour. This was "The Brain's Last Stand", according to the cover of *Newsweek*. If Deep Blue won, some articles suggested, humankind's self-image would undergo a transformation as dramatic as the adoption of the Copernican view of the universe, in which the earth was no longer at the centre, or Charles Darwin's theory of evolution, which explained how humans were descended from apes.

Playing white and opening the first game, Kasparov was firm and confident, and he gradually established an advantage. Suddenly, Deep Blue unleashed an unexpected attack that disconcerted the spectators. But Kasparov remained in control. Although his forces were weaker in the endgame, his position was better, and he ultimately triumphed. Afterwards, as an exultant Kasparov took questions from the audience, the Deep Blue team's disappointment was clear. It looked as though Kasparov had the measure of his opponent and was heading for a spectacular victory.

In the second game, Kasparov played far more defensively, deliberately orchestrating a dense, closed position of the kind that computers generally find hard to analyse. But this strategy backfired. Although Kasparov never made a serious error, his play was too passive, and when it became clear that he had been subtly outmanoeuvred by the computer, he resigned. Suddenly, his ability to win the match looked uncertain. Press coverage intensified, and IBM's share price rose by more than four dollars. Subsequent analysis revealed that Kasparov had resigned in a position where he could have forced a draw.

Kasparov later wrote that his defeat in the second game "left a scar in my memory and prevented me from achieving my usual total concentration in the following games". The third game was finely balanced and ended in a draw. As he had after the first game, Kasparov went down to the auditorium to answer questions from the spectators. But rather than talk about the third game, he spent most of his time talking about Deep Blue's play in the second game. One particularly cunning move, he had concluded, could only have been made by a human; Kasparov insisted that he "knew" how computers played chess, and that the move in question would only have occurred to one of the world's best human players, and not to a mere machine. In a sense, this was a compliment to the Deep Blue team, since Kasparov had conceded that the computer had played in a way that was indistinguishable from a human; it had, at least for one move, passed the chess equivalent of the Turing test. But Kasparov seemed to be hinting that Deep Blue had cheated, and that its move had really been suggested by one of its expert human "trainers". The Deep Blue team furiously denied this suggestion.

The fourth game also ended in a draw. Technically, the score was now even, but Kasparov looked drained and beaten. After a two-day rest, his play in the fifth game lacked its usual passion, and the result was another draw. Everything now hinged on the sixth and final game. Playing black, Kasparov followed a standard opening. But at the seventh move he made an elementary error, transposing two moves so that his defence unfolded in the wrong order. Horrified, he went on to lose the game in less than an hour. Deep Blue had won the match. For the first time, the world chess champion had lost a match to a computer in standard tournament conditions.

At the press conference afterward, Kasparov defended himself by repeating his concerns about the second game and demanding full printouts showing Deep Blue's reasoning. IBM refused to provide them. Writing in *Time* magazine after his defeat, Kasparov called for a further rematch. "I also think IBM owes me, and all mankind, a rematch," he declared. "I hereby challenge IBM to a match of 10 games, 20 days long, to play every second day. I would like to have access in advance to the log of 10 Deep Blue games played with a neutral player or another computer in the presence of my representative."

At a speech to a computing conference in Oregon in 1999, Kasparov complained that "IBM had a duty and still has a moral obligation to give the chess world access to the printouts." Any world record in sports is always followed by a drug test, he noted, and scientists are expected to back up their claims with the appropriate data. On neither sporting nor scientific grounds, he implied, had IBM played fair. Furthermore, he noted, "IBM collected several scientific awards from universities or different kinds of scientific institutions without disclosing any data to prove that it was a real machine." Like many of the Turk's opponents, Kasparov suspected that what had been presented as a pure, chess-playing machine had actually been guided by a hidden human operator.

IBM, which had always dismissed all suggestions of foul play, would not agree to another match. It is not hard to see why: the company estimated that the publicity associated with defeating Kasparov had been worth $500 million in advertising for its computers, its share price reached an all-time high, and it decided to quit while it was ahead. In yet another parallel with the Turk's career, IBM subsequently claimed to have dismantled the original machine, though privileged visitors to the Thomas J. Watson Research Center have occasionally been allowed to play against a smaller version of it.

*

Here at last was the fulfilment of Kempelen's dream: a chess-playing machine that could defeat the world's best players. And it was implemented using computers, rather than conjuring. But was Deep Blue really a triumph of machine over man? Not exactly. Just as the Turk was closely scrutinized throughout its career by pamphleteers bent on unmasking it, Deep Blue was also the subject of much detailed analysis and commentary. In an uncanny echo of Philip Thicknesse's vituperative attack on the Turk, an American philosopher, John Searle, published an essay in the *New York Review of Books* refuting the notion that Deep Blue could be considered in any way intelligent. Searle pointed out that in addition to its vast number-crunching power, Deep Blue had been programmed with thousands of rules, or "tactical weighting factors", devised by human experts. "The real competition was not between Kasparov and the machine, but between Kasparov and a team of engineers and programmers," he concluded. Deep Blue, like the Turk, relied on an illusion: it appeared to be a thinking machine, but effectively it had several human experts hiding inside it.

Unlike the Turk, however, Deep Blue did not rely on deception. The engineers who built it were quite open about how their illusion worked, and never claimed that their creation was intelligent. "I never consider Deep Blue to be intelligent in any way," said Murray Campbell, one of its creators. "It's just an excellent problem solver in this very specific domain. It would have been nice if Deep Blue had been able to teach itself to play, and learn from its own games, but it was painstakingly programmed every step of the way." In addition to its vast database of rules, Deep

Blue's ability to play chess depended on computational brute force, rather than an innate ability to analyse chess positions in the highly selective way a human expert can. The more processing power a computer has, the more chess positions and rules it can consider. In the case of Deep Blue, as Kasparov so accurately put it, "quantity had become quality".

Yet despite the relentless improvement in computer performance, there has been little progress toward general-purpose machine intelligence since the pioneers of artificial intelligence first started thinking about chess-playing machines. Deep Blue showed that building a computer capable of playing chess as well as any human reveals very little about intelligence in general, contrary to what was believed in the 1950s. But what of Turing's other yardstick, the ability to hold conversations?

There are now several programs that can pass the Turing test, at least some of the time. These conversation-holding programs have been tested in Internet chat rooms on unsuspecting humans, some of whom never realized that they were chatting to a machine, not a person. Of course, an anonymous Internet chat room is different from a pure Turing test, in which the human participants are expected to do their best to trip up the machines by asking, for example, where they last went on vacation. A more rigorous test is the annual competition established in 1990 by Hugh Loebner, a New York philanthropist. Loebner has offered a prize of $100,000 for the first program that can consistently convince people that it is human. Each year a handful of competing programs and a few human decoys are interrogated by a panel of expert judges, including philosophers, linguists, and psychologists. While the programs are good at small talk in Internet chat rooms, they are not so convincing under such close scrutiny. So far the prize is unclaimed, but each year the success rate of the most convincing entrants increases. The winner of the 1998 and 1999 contests was a program called Albert One, created by Robby Garner, an American programmer. A typical conversation with Albert, who pretended to be a human named Jim Lewis, went as follows:

JUDGE: Hi, Jim. My name is Louise. Where do you live?

ALBERT: You're Louise, aren't you? I had a really hard time getting here today.

JUDGE: Where do you live, Jim?

ALBERT: Let's not bother ourselves about where I do live, Jim, all right?

JUDGE: How did you get here today?

ALBERT: Somebody told me.

JUDGE: Where do you work?

ALBERT: I work in a bar.

JUDGE: In what town?

ALBERT: I'm from Georgia, you know.

JUDGE: Where do you want to go for holiday?

ALBERT: Sometimes I go to Kroger.

JUDGE: What is the weather like today?

ALBERT: It is usually quite mild.

JUDGE: What is it like today?

ALBERT: I like phasers.

JUDGE: Do you have any children?

ALBERT: I'm sure we have at least some facsimile of a replica or something.

JUDGE: Do you have any sisters or brothers?

ALBERT: No. I'm an only child.

In pure Turing-test conditions, today's best programs can fool humans only about 15 percent of the time after a five-minute conversation, which means they are correctly identified as nonhumans 85 percent of the time. In other words, Turing's prediction that this figure would fall to 70 percent by the year 2000 has not come to pass. Furthermore, as with Deep Blue, the programs in question are highly specialized; unlike HAL, whose intelligence could be applied to a wide range of tasks, conversation-holding programs are no good at doing anything else. (That said, it is worth pointing out that in many cases, humans who take part in the tests are wrongly accused of being computers.)

Might more powerful computers make genuine machine intelligence possible in the future? Some researchers believe intelligence, like chess, is a matter of quantity, not quality, and that it is simply a question of having enough processing power. In the 1960s, this was known as the "wait until 2000" argument; the lack of progress in general-purpose machine intelligence was attributed to the lack of fast-enough hardware. This view is still championed by some researchers today, including the American software guru Raymond Kurzweil, who believes that superintelligent machines are just around the corner. In his 1999 book *The Age of Spiritual Machines: When Computers Exceed Human Intelligence,* he predicts that following current trends, by 2020 a $1,000 computer will have the same raw processing power as a human brain, which he estimates to be 20 million billion calculations per second. By 2030, Kurzweil suggests, the software will somehow have been developed to allow this hardware to match the flexibility and intelligence of the human brain, at which point it will supposedly be possible for people to upload themselves into computers and become immortal. Similar forecasts have been made by Hans Moravec of Carnegie Mellon University. But Kurzweil and Moravec's essentially numerical arguments about the inevitability of machine intelligence (based on the number of calculations per second) are strongly reminiscent of the flawed nineteenth-century speculations about the impossibility of building chess-playing machines (based on the total number of possible chess positions). There is surely more to creating an intelligent machine than simply building a faster computer.

Even so, it will probably become possible to build specialized systems, such as speech-controlled computers, that appear to be intelligent, just as Deep Blue appears to understand how to play chess. As with the Turk, the users of such machines might not always be able to tell whether they are dealing with a human or a machine. From a philosophical point of view, such machines will rely on trickery; they will not be truly intelligent. But Turing's point was that in practical terms, the illusion of intelligence is as good as the real thing. The Turing test thus remains the most enduring yardstick for machine intelligence. Amazingly, this is something that Wolfgang von Kempelen seems to have foreseen.

*

In the spring of 1770, when Kempelen first demonstrated the Turk to the astonished Viennese court, he opened the doors of the main compartment and took out three items: a red cushion, which went under the automaton's left arm; a small wooden casket, which he peered into during the performance; and a board marked with gold letters, which he placed on a nearby table. Once he had finished demonstrating the Turk's chess-playing abilities, Kempelen took this board and placed it on top of the Turk's chessboard. He then invited the audience to pose questions to the automaton. It replied by indicating letters on the board with its left hand, one at a time, to spell out its answers. Was it a human or a machine? The audience was baffled. The Turk, in other words, could pass the Turing test.

Such question-and-answer sessions, like the Knight's Tour, were merely an optional element of the Turk's performance, and Maelzel discontinued the practice since he thought that it undermined the illusion. For while audiences might be prepared to believe in the possibility of a mechanical chess player, a mechanical device that could understand speech and reply by spelling out its answers was plainly impossible. To a modern observer, the Turk's ability to answer questions is a dead giveaway that it was controlled by a human, and even at the time, the fact that the Turk answered in French while in Paris and in German while in Leipzig was regarded as suspicious. That said, it seems that many observers believed the Turk's chess playing to be genuinely machine-driven, while a human operator enabled it to respond to spoken questions via an entirely separate conjuring trick.

In any case, it is clear that in 1769 Kempelen had conjectured that playing chess and holding conversations were the two activities that most readily indicated intelligence. Nearly 200 years later, the computer scientists of the twentieth century came to exactly the same conclusion. It is ironic that the Turing test relies on concealment and deception: on machines trying to act like people, and people acting like machines. Kempelen would surely have been surprised to discover that, in a sense, little has changed since he unveiled his mechanical Turk. He regarded the automaton as an amusement and tried hard to prevent it from upstaging his other, more serious, accomplishments. But by embodying Kempelen's pioneering insight into the curious relationship between technology and trickery, intelligence and illusion, the Turk proved to be his greatest and most farsighted creation. The wily automaton has had the last laugh after all.

Notes

CHAPTER 1. *The Queen's Gambit Accepted*

The descriptions of eighteenth-century automata are drawn from Bedini, "The Role of Automata"; Chapuis and Droz, *Automata;* Schaffer, "Babbage's Dancer"; and Altick, *The Shows of London.* (Altick provided the story of the false harpsichord player.) Vaucanson's biographical details are drawn from Doyon and Liaigre, *Jacques Vaucanson,* and Strauss, "Automata." Maria Theresa's quotation about witchcraft is from Morris, *Maria-Theresa.* Kempelen's biographical details, and the description of his challenge and of his workshop, are drawn from Windisch, *Inanimate Reason;* Würzbach, "Kempelen"; and personal correspondence with Alice Reininger and Viktoria Schurk, who provided information from additional German and Hungarian sources. Kempelen is widely but wrongly described as a baron in many sources. He was, however, merely a minor noble who was entitled to use the German honorific *von* in front of his surname (translated as *de* in French). Some sources describe Kempelen as a *Ritter,* or knight, but there is no evidence that he was ever granted such a title. It is particularly noteworthy that the frontispiece of Kempelen's book, published near the end of his life in 1791, does not refer to him as either a baron or a knight. Kempelen's first name is sometimes given as Farkas, the Hungarian for wolf. But the name on his birth certificate is Wolfgang.

CHAPTER 2. *The Turk's Opening Move*

The description of the Turk's appearance and performance is drawn from Windisch, *Inanimate Reason,* and Racknitz, *Ueber den Schachspieler.* The Knight's Tour pattern followed by the Turk can be found in Levitt, *The Turk;* the original pattern was last seen at the Library Company of Philadelphia but seems to have disappeared. Dutens's letters to *Le Mercure de France* are quoted from Dutens, "Lettres sur un automate"; the English translation follows that of the *Gentleman's Magazine* where appropriate. Kempelen's decision to dismantle the Turk is recounted by Windisch, *Inanimate Reason.* Sir Robert Murray Keith's account is quoted from his *Memoirs and Correspondence;* Kempelen's letter is quoted from the original manuscript at the British Library in London.

CHAPTER 3. *A Most Charming Contraption*

Kempelen's reluctant restoration of the Turk at the behest of Joseph II is recounted by Windisch, *Inanimate Reason.* Paris and London as centres of enthusiasm for chess are described by Eales, *Chess.* The quote about Legall, and biographical information about Philidor, is from Hooper and Whyld, *The Oxford Companion to Chess.* The Turk's adventures in France are described by Bachaumont, *Mémoires secrets;* Grimm, *Correspondence;* and Croy, *Mémoires.* Valltravers's letter to Franklin is quoted from Ewart, *Chess.* Kempelen's letter to Franklin is quoted from Evans, *Edgar Allan Poe.* The example of Philidor's Legacy is quoted from Pritchard, *The Right Way to Play Chess.* Philidor's supposed agreement to lose a game to the Turk is recounted in *Le Palamède* 7 (1847), 12–13. The account of the French savants' attempt to

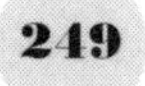

figure out how the Turk works appears in the *Journal des Savants* of September 1783, as part of a review of Windisch's book.

CHAPTER 4. *Ingenious Devices, Invisible Powers*

The vogue for automata in eighteenth-century London is described by Altick, *The Shows of London,* and Schaffer, "Babbage's Dancer". The Johnson quotation comes from Altick's book. The Brewster quotation is from Schaffer's article. Cartwright's story is quoted from Altick's book.

CHAPTER 5. *Dreams of Speech and Reason*

Böckmann's theory is quoted from Carroll, *The Great Chess Automaton.* The information about Kempelen's speaking machines comes from Windisch, *Inanimate Reason;* Flanagan, "The Synthesis of Speech"; and Chapius and Droz, *Automata.* Bell's interest in Kempelen's research is detailed in Mackay, *Sounds and Silence.* Details of Collinson's visit to Kempelen are quoted from Carroll's book.

CHAPTER 6. *Adventures of the Imagination*

The Frederick the Great tale is recounted in Walker, "Anatomy of the Chess Automaton", among other sources. The Catherine the Great story is told in Robert-Houdin, *Memoirs.* Details of the books and plays inspired by the Turk can be found in Ewart, *Chess.* Babbage's interest in automata is detailed in his autobiography, *Passages.* Details of Kempelen's later life are taken from Würzbach, "Kempelen", and personal correspondence with Alice Reininger and Viktoria Schurk, who provided information from additional German and Hungarian sources. After Kempelen's death his biography seems to have been reinvented by Hungarian nationalists seeking independence from Austria, who claimed that Kempelen was stripped of all his assets by Francis II and died penniless. But according to Alice Reininger, Kempelen continued to receive his pension right up to his death.

CHAPTER 7. *The Emperor and the Prince*

The account of Napoleon's encounter with the Turk is from Wairy (Constant), *Mémoires.* Maelzel's biographical details are from Würzbach, "Maelzel"; Fétis, "Maelzel"; and Thayer, *Thayer's Life of Beethoven.* Biographical information about Eugène de Beauharnais comes from Oman, *Napoleon's Viceroy,* and Du Casse, *Mémoires.* Napoleon's interest in chess is detailed in Ewart, *Chess.* Eugène's purchase of the Turk, and the sightings of the automaton in Milan, are described in George Allen, "The History of the Automaton Chess Player". The letter from Maelzel referring to the Turk as "the automaton chess-player entrusted to me by Prince Eugène" is quoted from Chapuis and Droz, *Automata.*

CHAPTER 8. *The Province of Intellect*

The epigraph is quoted from Ewart, *Chess.* Willis's biographical details are from Stephen, ed., "Willis, Robert". Babbage's inspiration for the Difference Engine is detailed in his autobiography, *Passages.* His encounters with the Turk were described on a piece of paper found inside his copy of Windisch's book, *Inanimate Reason;* this book is now in the British Library, and Babbage's description is quoted in "De Kempelen's Automaton Chess-Player", in *Notes and Queries.* Babbage's interest in the prospect of machine intelligence, and his party trick with the Difference Engine, are described by Schaffer, "Babbage's Dancer." Lewis's comments about Maelzel are quoted from Ewart, *Chess.* The building on St. James's Street where the Turk was displayed is just up the road from the *Economist*'s modern-day offices.

CHAPTER 9. *The Wooden Warrior in America*

The main source for the Turk's adventures in America is George Allen, "The History of the Automaton Chess Player", with additional material (including newspaper quotations) from Wittenberg, "Échec!"; Ewart, *Chess;* and Carroll, *The Great Chess Automaton.* The Barnum quotation comes from Altick, *The Shows of London.*

CHAPTER 10. *Endgame*

Krutch's quote about Poe is from Evans, *Edgar Allan Poe;* Hervey Allen's is from his Poe biography, *Israfel.* Maelzel's last days, and those of the Turk, are detailed in George Allen, "The History of the Automaton Chess Player". The purchase and reconstruction of the Turk are described by Mitchell, "Last of a Veteran Chess Player".

CHAPTER 11. *The Secrets of the Turk*

The explanation of the Turk's mechanism is based on the accounts in Mitchell, "Last of a Veteran Chess Player", and in Smith's manuscript. The details of the Turk's various operators are from Ewart, *Chess.* The story of William Coleman's son sitting in as the Turk's operator in place of the young Frenchwoman is told in George Allen, "The History of the Automaton Chess Player". The stories of the conjuror who yelled "Fire!" and the Turk's supposed game with the king of Holland are from Walker, "Anatomy of the Chess Automaton". Davidson's incorrect theory about the Turk is quoted from Levitt, *The Turk.* The account of Gaughan's reconstruction of the Turk is based on interviews with Gaughan in London and Los Angeles. Gümpel's and Proctor's theories are described in Ewart, *Chess.*

CHAPTER 12. *The Turk Versus Deep Blue*

Glennie's account of playing against Turing's paper machine is quoted from Bell, *The Machine Plays Chess.* The brief history of computer chess is drawn from Campbell, "An Enjoyable Game". The development of Deep Thought and Deep Blue is described by Campbell in the same source, and by documents on IBM's Web site (www.ibm.com). The account of Kasparov's matches against Deep Blue is from Schaeffer and Plaat, "Kasparov Versus Deep Blue". Kasparov's side of the story is based on transcripts of his speeches provided by his agent, Owen Williams. The quote from Murray Campbell is from a telephone interview with the author conducted in 1998. For more information about the Loebner Prize, see www.loebner.net. The transcript of the conversation with Albert is from Pescovitz, "How to Send a Bot Off on a Rant?" The most detailed accounts of the Turk's ability to answer questions are provided by Ebert, *Nachricht,* and Hindenburg, "Ueber den Schachspieler", and are quoted in Carroll, *The Great Chess Automaton,* among other sources.

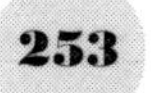

Sources

Allen, George. "The History of the Automaton Chess Player in America." In *The Book of the First American Chess Congress.* New York: Rudd and Carleton, 1859, 420–84.

Allen, Hervey. *Israfel: The Life and Times of Edgar Allan Poe.* New York: George H. Doran, 1927.

Altick, Richard. *The Shows of London.* Cambridge, Mass.: Harvard University Press, 1978.

"Automate jouer d'échecs." *Magazine Pittoresque* 2 no. 20 (1834): 155.

"The Automaton Chess Player Redivivus." *Illustrated London News,* December 20, 1845, 389–90.

Babbage, Charles. *Passages from the Life of a Philosopher.* London: Longman, 1864.

Bachaumont, Louis Petit de. *Mémoires secrets pour servir à l'histoire de la republique des lettres en France.* Vol. 22, 249–50, 262–64, 305–7; vol. 23, 3–6 (April–June 1783). London, 1777–89.

Bedini, Silvio: "The Role of Automata in the History of Technology." *Technology and Culture* 5 (1964): 24–42.

Bell, Alex G. *The Machine Plays Chess.* Oxford: Pergamon, 1978.

Bourdonnais, Charles de la. "L'Automate jouer d'échecs." *Le Palamède* 4 no. 2–3 (1839): 54–70.

Sources

Bradford, Gamaliel. *The History and Analysis of the Supposed Automaton Chess Player of M. de Kempelen*. Boston: Hilliard, Gray, 1826.

Brewster, Sir David. *Letters on Natural Magic*. London: John Murray, 1832.

Campbell, Murray S. "An Enjoyable Game: How HAL Plays Chess." In *HAL's Legacy: 2001's Computer as Dream and Reality*, ed. David G. Stork. Cambridge, Mass.: MIT Press, 1997, 74–98.

Carroll, Charles Michael. *The Great Chess Automaton*. New York: Dover Publications, 1975.

Chapuis, Alfred, and Droz, Edmund. *Automata*. London: Batsford, 1958.

Clarke, Arthur C. *2001: A Space Odyssey*. London: Hutchinson, 1968.

Croy, Duc de. *Mémoires sur les cours de Louis XV et de Louis XVI*. Paris, 1895–96.

"De Kempelen's Automaton Chess-Player." *Notes and Queries* 12 S.X. February 25, 1922, 155–56.

De Tournay, Mathieu-Jean-Baptiste Nioche. "La Vie et les aventures de l'automate jouer d'échecs." *Le Palamède* 1, no. 3 (1836): 81–87.

Decremps, Henri. *La Magie blanche devoilée*. Paris: Langlois, 1784.

Doyon, André, and Liaigre, Lucien. *Jacques Vaucanson, mécanician de génie*. Paris: PUF, 1966.

Du Casse, André, ed. *Mémoires et correspondence politique et militaire du Prince Eugène*. Paris, Michel Lévy, 1859.

Dutens, Louis. "A Description of an Automaton, Which Plays at Chess." *Gentleman's Magazine*, January 1771, 26–27.

———. "Lettres sur un automate, qui joue aux échecs." In *Oeuvres mélees, 103–9*. London, P. Emsley, 1796.

Eales, Richard. *Chess: The History of a Game*. London: Batsford, 1985.

Ebert, Johann Jakob. *Nachricht von dem berühmten Schachspieler und der Sprachmaschine des Herrn von Kempelen*. Leipzig, 1785.

Encyclopaedia Britannica. 11th ed. "Automaton" and "Conjuring."Cambridge: Cambridge University Press, 1911.

Evans, Henry Ridgely. *Edgar Allan Poe and Baron von Kempelen's Chess-Playing Automaton.* Kenton, Ohio: International Brotherhood of Magicians, 1939.

Ewart, Bradley. *Chess: Man vs Machine.* London: Tantivy Press, 1980.

Fétis, F. J. "Maelzel." In *Biographie universelle des musiciens,* 396–97. Paris, 1870.

Flanagan, James L. "The Synthesis of Speech." *Scientific American,* February 1972, 48–58.

Grimm, Frédéric Melchior. *Correspondence littéraire, philosophique, et critique.* Vol. 13 (September 1783). Paris: Chez Furne, 1830.

Hindenburg, Carl Friedrich. "Ueber den Schachspieler des Herrn von Kempelen, nebst einer Abbildung und Beschreibung seiner Sprachmaschine." *Leipziger Magazin zur Naturkunde, Mathematik, und Oekonomie,* 235–69. Leipzig, 1784.

Hooper, David, and Whyld, Ken. *The Oxford Companion to Chess.* Oxford: Oxford University Press, 1992.

Hunneman, W. *A Selection of Fifty Games, from Those Played by the Automaton Chess-Player, During Its Exhibition in London in 1820.* London, 1820.

"Inanimate Reason" (review). *Monthly Review* 70 (April 1784): 307–8.

Jarrett, Keith. *Whisper Not.* Munich: ECM Records, 2000.

Keith, Sir Robert Murray. *Memoirs and Correspondence, Official and Familiar, Edited by Mrs Gillespie Smyth.* London: Colburn, 1849.

Kempelen, Wolfgang von. Manuscript letter to Robert Murray Keith, August 14, 1774. British Library manuscript collection ref. 355 07, vol. 5, 275.

———. *Mechanismus der menschlichen Sprache nebst Beschreibung einer sprechenden Maschine.* Vienna: J. B. Degen, 1791.

Knudsen, John C. *Essential Chess Quotations.* Osthofen: John C. Knudsen, 1998.

Köszega, Imre, and Pap, János. *Kempelen Farkas.* Budapest, 1955.

Sources

Kurzweil, Raymond. *The Age of Spritual Machines: When Computers Exceed Human Intelligence.* London: Orion, 1999.

"Lettres sur le jouer d'échecs de M. de Kempelen" (review). *Journal des Savants* (September 1783.)

Levitt, Gerald M. *The Turk, Chess Automaton.* Jefferson, N.C.: McFarland, 2000.

Lopez, Claude-Anne. *Mon cher papa: Franklin and the Ladies of Paris.* New Haven: Yale University Press, 1966.

Mackay, James. *Sounds out of Silence: A Life of Alexander Graham Bell.* Edinburgh and London: Mainstream Publishing, 1997.

Mitchell, Silas Weir. "Last of a Veteran Chess Player." *Chess Monthly,* January 1857, 3–7, and February 1857, 40–45.

Morris, Constance Lily. *Maria-Theresa, the Last Conservative.* London: Eyre and Spottiswoode, 1938.

Observations on the Automaton Chess Player Now Exhibited in London by an Oxford Graduate. London: J. Hatchard, 1819.

Oman, Carola. *Napoleon's Viceroy, Eugène de Beauharnais.* London: Hodder and Stoughton, 1966.

Pérez-Reverte, Arturo. *The Flanders Panel.* London: Harvill Press, 1994.

Pescovitz, David. "How to Send a Bot off on a Rant? Mention Star Trek." *New York Times,* March 18, 1999.

Poe, Edgar Allan. *The Complete Tales and Poems of Edgar Allan Poe.* London: Penguin Books, 1965.

Porter, Roy. *The Greatest Benefit to Mankind: A Medical History of Humanity from Antiquity to the Present.* London: HarperCollins, 1997.

———, ed. *Hutchinson Dictionary of Scientific Biography.* Oxford: Helicon, 1994.

Pritchard, D. Brine. *The Right Way to Play Chess.* 10th ed. Kingswood, Surrey: Elliot Right Way Books, 1974.

Racknitz, Joseph Friedrich, Freyherr zu. *Ueber den Schachspieler des*

Herrn von Kempelen und dessen Nachbildung. Leipzig and Dresden: verlegts Johann Gottlieb Immanuel Breitkopf, 1789.

Robert-Houdin, Jean Eugène. *Memoirs of Robert-Houdin, Ambassador, Author, and Conjurer, Written by Himself.* London: Chapman and Hall, 1859.

Schaeffer, Jonathan, and Plaat, Aske. "Kasparov Versus Deep Blue: The Re-match." *ICCA Journal* 20, no. 2 (1997): 95–102.

Schaffer, Simon. "Babbage's Dancer and the Impresarios of Mechanism." In *Cultural Babbage,* ed. Francis Spufford and Jenny Uglow. London: Faber and Faber, 1996.

Schurk, Viktoria. "Ungarische Erfinder." *Budapester Zeitung,* September 4, 2000.

"Scientific Amusements – Automata." *New Monthly Magazine* 1 (1821): 441–48, 524–32.

Searle, John. "I Married a Computer." *New York Review of Books,* April 8, 1999.

Smith, Lloyd Pearsall. Manuscript letter to George Allen, July 7, 1858. Library Company of Philadelphia, Allen (Chess) Collection ref. Yi2 7425 F. 31.

Stephen, Leslie, ed. "Willis, Robert." In *Dictionary of National Biography.* London: Smith, Elder, 1885–1903.

Strauss, Linda M. *"Automata."* Ph.D. thesis, London Science Museum Library ref. 688:93 STRAUSS.

Thayer, Alexander Wheelock. *Thayer's Life of Beethoven, Revised and Edited by Elliot Forbes.* Princeton: Princeton University Press, 1970.

Thicknesse, Philip. *The Speaking Figure, and the Automaton Chess-Player, Exposed and Detected.* London, 1784.

Turing, Alan. "Computing Machinery and Intelligence." *Mind* 59, no. 236 (October 1950): 433–60.

Twiss, Richard. *Chess.* London: G.G.I. and I. Robinson, 1787.

Wairy, Louis Constant. *Mémoires de Constant, premier valet de chambre de l'empereur sur la vie privée de Napoléon, sa famille, et sa cour.* Paris, 1830.

Walker, George. "Anatomy of the Chess Automaton." *Fraser's Magazine* 19 (June 1839): 717–31.

Whyld, Ken. *Fake Automata in Chess* (bibliography). Lincoln, England: Caistor, 1994.

———. "Maelzel's Little Book." *British Chess Magazine* 120, no. 7 (July 2000): 382–84.

Willis, Robert. *An Attempt to Analyse the Automaton Chess Player, of Mr de Kempelen. With an Easy Method of Imitating the Movements of That Celebrated Figure.* London: J. Booth, 1821.

Windisch, Carl Gottlieb von. *Inanimate Reason; or a Circumstantial Account of That Astonishing Piece of Mechanism, M. de Kempelen's Chess-Player.* London, 1784.

Wittenberg, Ernest. "Échec! The Bizarre Career of the Turk." *American Heritage,* February 1960: 34–37, 82–85.

Würzbach, Constant von. "Kempelen, Wolfgang Ritter von" and "Maelzel, Johann Nepomuk." In *Biographisches Lexikon des Kaiserthums Oesterreich.* 60 vols. Vienna, 1856–91.

Acknowledgments

There turns out to be something of a "Turk mafia" – a group of Turk enthusiasts around the world that includes scholars, chess enthusiasts, computer scientists, and historians of magic – many of whom provided me with invaluable assistance. I am particularly grateful to Alice Reininger in Vienna and Viktoria Schurk in Budapest, who provided me with access to archival material and German and Hungarian sources that would otherwise have been inaccessible. I would also like to thank Ken Whyld for his indispensable bibliography, Bill Kuethe for plugging me into the Turk mafia in the first place, and Gerald Levitt, whose book on the Turk is an essential resource for anyone interested in the subject. Special thanks go to John Gaughan, who not only invited me to Los Angeles to see his reconstructed Turk but generously shared the many insights into Kempelen's automaton that he gained in the process of building his own version. My thanks are also due to Leonard Barden, Simon Schaffer (to whom I now owe three pints of beer), Oliver Morton, Sally Forbes, Owen Williams, Vendeline von Bredow, and Demetrio Carrasco; and to Chester for both acting as an "intelligent Martian" and providing the Keith Jarrett soundtrack to which much of this book was written. Thanks too to Katinka Matson, George Gibson, Jackie Johnson, Cassie Dendurent, Stefan McGrath,

Maria Iosifescu, Virginia Benz and Joe Anderer, Anna Aebi, Sue Docherty, Tamsin Murray-Leach, Lee McKee, and Tom and Kathryn Moultrie. Finally, I would like to thank my wife, Kirstin, for, well, everything really.

Index

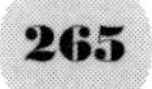

Index

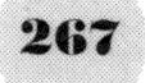

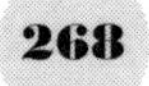

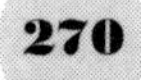

Index

Index